Two From One
Jelly Roll Quilts

Two From One Jelly Roll Quilts

18 designs to make your fabric go further

PAM AND NICKY LINTOTT

David and Charles
www.sewandso.co.uk

A DAVID & CHARLES BOOK
© F&W Media International, Ltd 2010

David & Charles is an imprint of F&W Media International, Ltd
Pynes Hill Court, Pynes Hill, Exeter, EX2 5AZ, UK

F&W Media International, Ltd is a subsidiary of F+W Media, Inc
10151 Carver Road, Suite #200, Blue Ash, OH 45242, USA

First published in the UK and US in 2010
Reprinted in 2010 (twice), 2011, 2015, 2016

Text and designs © Pam Lintott and Nicky Lintott 2010
Layout and photography © David & Charles 2010

Pam Lintott and Nicky Lintott have asserted their right to be identified as authors of this work in accordance with the Copyright, Designs and Patents Act, 1988.

All rights reserved. No part of this publication may be reproduced, stored in a retrieval system, or transmitted, in any form or by any means, electronic or mechanical, by photocopying, recording or otherwise, without prior permission in writing from the publisher.

The designs in this book are copyright and must not be stitched for resale.

The author and publisher have made every effort to ensure that all the instructions in the book are accurate and safe, and therefore cannot accept liability for any resulting injury, damage or loss to persons or property, however it may arise.

Names of manufacturers, fabric ranges and other products are provided for the information of readers, with no intention to infringe copyright or trademarks.

A catalogue record for this book is available from the British Library.

ISBN-13: 978-0-7153-3756-1 paperback
ISBN-10: 0-7153-3756-4 paperback
ISBN-13: 978-1-4463-1246-9 hardcover
ISBN-10: 1-4463-1246-1 hardcover

F&W Media International, Ltd
Pynes Hill Court, Pynes Hill, Exeter, EX2 5AZ, UK

Commissioning Editors Jane Trollope and Cheryl Brown
Assistant Editor Juliet Lines
Project Editor Lin Clements
Design Manager Sarah Clark
Photographers Sian Irvine and Karl Adamson
Production Manager Bev Richardson
Pre Press Jodie Culpin

David and Charles publish high quality books on a wide range of subjects.
For more great book ideas visit: www.sewandso.co.uk

Contents

Introduction 6
Getting Started 8

The Quilts

Chapter One 12
Building Blocks Quilt 14
Sherbert Lemon Quilt 17

Chapter Two 20
Speedwell Quilt 22
Bubblegum Quilt 26

Chapter Three 30
Square Dance Quilt 32
Strip the Willow Quilt 35

Chapter Four 38
Sailing Boats Quilt 40
Roller Coaster Quilt 44

Chapter Five 48
Stardust Quilt 50
Playmates Quilt 54

Chapter Six 58
Snapdragon Quilt 60
Fairy Steps Quilt 64

Chapter Seven 68
Spinning Tops Quilt 70
Train Tracks Quilt 74

Chapter Eight 80
Kaleidoscope Quilt 82
Jigsaw Quilt 88

Chapter Nine 94
Teddy Bears Quilt 96
Loving Hearts Quilt 102

General Techniques

General Techniques 110
Useful Information 124
Useful Contacts 125
Acknowledgments 126
About the Authors 126
Index 128

Introduction

The basis of all the quilts in this book is just *half* a jelly roll, so from one of those delicious jelly rolls you can make *two* jelly roll quilts. Yes, all of the lovely quilts in this book are made with only twenty jelly roll strips.

Some of the quilts require a little sorting of colours and these quilts are always featured first in the chapter. Once your choice of strips for the first quilt is made then your second quilt is made from the twenty strips remaining. How great is that, and see how different they look, both in design and colour. Two for the price of one – how economical!

If you don't have a jelly roll, you can always cut your own twenty 2½in strips from your fabric stash to make any of the quilts in this book. However you manage to put together your twenty strips you now have no excuse for not making those cot quilts, play mats, throws and wall hangings for your loved ones. There are lots of exciting designs for you to choose from, all of them quick and easy to make, so off you go and just have fun.

If you want to use any of the designs to make a larger quilt, refer to the vital statistics panel at the start of the quilt instructions – all the information you need is there and with a few small calculations you can make any size quilt you want. You could even use one design for the front and one for the back to make a reversible quilt – there are lots of exciting possibilities!

Don't worry if you can't find the same jelly rolls as we have used. Remember, fabrics are just like clothes – they come in for one season and are no longer available the next. So just be guided by what you have in the jelly roll you have chosen and you will end up with a quilt that is unique to you. Have fun!

Getting Started

What is a Jelly Roll?
Moda introduced jelly rolls to showcase new fabric ranges and a jelly roll is a roll of forty fabrics cut in 2½in wide strips across the width of the fabric. How inspirational to have one 2½in wide strip of each new fabric wrapped up so deliciously! Our thanks go to Moda for inspiring us and allowing us to use the name jelly roll in this book. All the quilt patterns in this book assume that fabric will be at least 42in wide.

Two From One Jelly Roll Quilts uses one jelly roll to make up two different quilts. Yes, I know we should be content with making one quilt from one jelly roll but hey, let's make the most of these gorgeous strips of fabric. If you want to make any of the quilts in this book and don't have a jelly roll to use, then cut a 2½in wide strip from twenty fabrics in your stash and you can follow all the instructions in just the same way to make one quilt.

Imperial or Metric?

Jelly rolls from Moda are cut 2½in wide and at The Quilt Room we have continued to cut our strip bundles 2½in wide. When quilt making, it is impossible to mix metric and imperial measurements. It would be absurd to have a 2½in strip and tell you to cut it 6cm to make a square! It wouldn't be square and nothing would fit. This caused a dilemma when writing instructions for our quilts and a decision had to be made. All our instructions therefore are written in inches. To convert inches to centimetres, multiply the inch measurement by 2.54 – see also the handy conversion chart on page 124. For your convenience, the fabric you will need, given in the Requirements panel at the start of the quilt instructions, are given in both metric and imperial.

Seam Allowance

We cannot stress enough the importance of maintaining an accurate scant ¼in seam allowance throughout. Please take the time to check your seam allowance with the test on page 112.

Quilt Size

In this book we have shown what can be achieved with just half a jelly roll (twenty 2½in strips). We have sometimes added background fabric and borders but the basis of each quilt is just half a jelly roll. Refer to the Vital Statistics panel at the start of the quilt instructions to calculate how many strips you need to make a larger quilt.

Diagrams

Diagrams have been provided to assist you in making the quilts and these are normally beneath or beside the relevant stepped instruction. The direction in which fabric should be pressed is indicated by arrows on the diagrams. The reverse side of the fabric is shown in a lighter colour than the right side. Read all the instructions through before starting work on a quilt.

Washing Notes

It is important that pre-cut strips are ***not*** washed before use. Save the washing until your quilt is complete and then use a colour catcher in the wash or possibly dry clean the quilt.

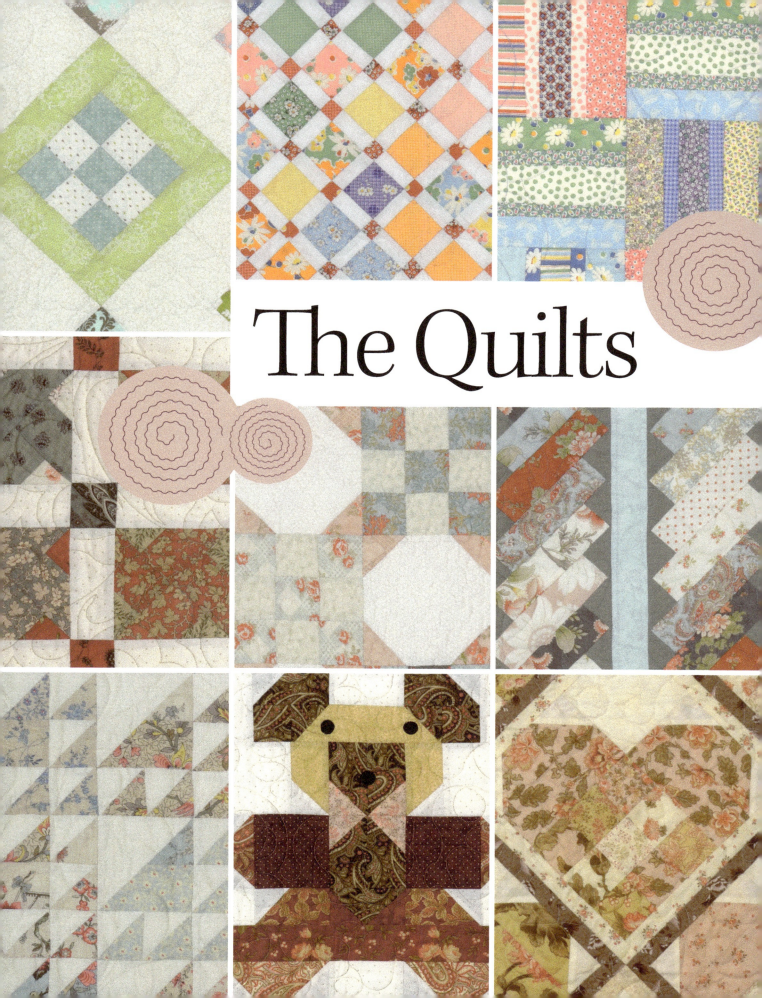
The Quilts

Chapter One
Building Blocks and Sherbert Lemon

We used a lovely pastel jelly roll for these two charming cot quilts and very little sorting of strips was needed. Both quilts are quick and easy to make and you could create one for a nursery in no time at all.

For the Building Blocks quilt we could have been more daring with our accent fabric but we loved this fabric and couldn't resist using it. Be aware that if the fabric is too light you will lose the effect of the pattern.

The Sherbert Lemon quilt showcases the four-patch block beautifully. It is such a simple block but look how effective it is when set on point with the rows of stripes in between.

Building Blocks Quilt

VITAL STATISTICS
Quilt size: 46in x 46in
Block size: 6in square
Number of blocks: 9
Setting: 3 x 3 plus connector blocks

REQUIREMENTS
- Half a jelly roll OR twenty 2½in wide strips cut across the width of the fabric
- 16in (40cm) of fabric for background
- 28in (70cm) of fabric for accent and border
- 16in (40cm) of fabric for binding

Sorting the Strips
- Choose eight strips for the thirteen nine-patch blocks.
- Choose nine strips to be the outer strips for nine of the nine-patch blocks. Each jelly roll strip will be the outer strip for one nine-patch block. The remaining three jelly roll strips are spare.

Cutting Instructions
Jelly roll strips:
- Cut each of the nine jelly roll strips to be used as the outer strips for the nine-patch blocks into the following, keeping the rectangles from each strip together:
cut two rectangles each 2½in x 6½in;
cut two rectangles each 2½in x 10½in.

Background fabric:
- Cut two 6½in wide strips across the width of the fabric.

Accent and border fabric:
- Cut nine 2½in wide strips across the width of the fabric. Set five strips aside for the border. The remaining four will be used to make the connector blocks.

Binding fabric:
- Cut five 2½in wide strips across the width of the fabric.

Making the Nine-Patch Blocks

1. Pair up the eight jelly roll strips allocated for the nine-patch blocks. Take one pair of strips and cut each into three lengths of 14in.

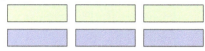

2. From your six lengths sew two strip units. Press to the darker fabric. Cut each strip unit into five 2½in segments.

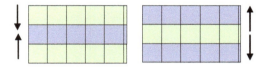

3. Make three nine-patch blocks and set aside the extra segment. Press the work. Repeat with the other three sets of strips. You will have twelve nine-patch blocks. Make up a thirteenth nine-patch block from three of the extra segments. One segment will be spare.

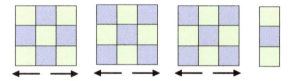

Note: If you want a scrappier effect for the nine-patch blocks, cut the eight strips allocated for the nine-patch blocks into 14in lengths and then join the strips randomly to make the strip units.

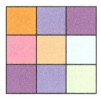

Adding the Outer Strips

4. Sew a 2½in x 6½in strip of the same fabric to the top and bottom of the nine-patch block. Press the seams away from the nine-patch block.

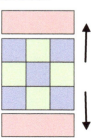

Be as bold as you dare with your accent fabric for this cot quilt to create the wonderful three-dimensional effect with the squares. What a lovely gift this would make for a couple expecting their first child. The quilt was pieced by the authors and longarm quilted by The Quilt Room.

Building Blocks Quilt

5 Sew a 2½in x 10½in strip of the same fabric to either side of this unit. Press the seams away from the nine-patch block.

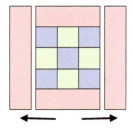

6 Repeat to make nine nine-patch blocks with outer strips. Leave the remaining four nine-patch blocks without outer strips.

Sewing the Connector Blocks

7 Sew a 2½in accent strip to either side of a 6½in background strip. Press towards the background fabric. Cut into six 6½in segments. Repeat with the other 6½in background strip and accent strips to make twelve connector blocks in total.

Assembling the Quilt Top

8 Referring to the diagram below, lay out the nine-patch blocks with outer strips, the connector blocks and the nine-patch blocks without outer strips. When you are happy with the layout sew the blocks together into rows. Press as shown in the diagram below, always in the direction of the connector blocks. Sew the rows together pinning at every seam intersection to ensure a perfect match.

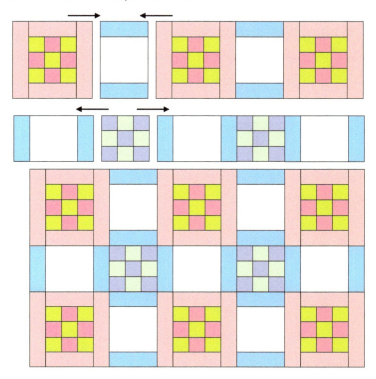

Adding the Borders and Finishing

9 Join the five 2½in border strips into one continuous length and refer to the instructions on page 117 to add the borders to the quilt.

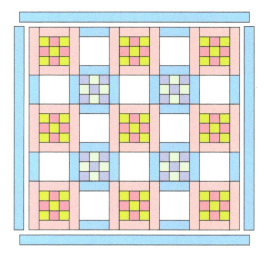

10 Your quilt top is now complete. Quilt as desired (see page 118 for advice) and bind to finish (page 118).

16   Building Blocks Quilt

Sherbert Lemon Quilt

VITAL STATISTICS
Quilt size: 35in x 53in
Block size: 6in square
Number of blocks: 16 plus three strip rows
Setting: two rows of eight blocks plus three 6in wide strip rows and a 2½in border

REQUIREMENTS
- Half a jelly roll OR twenty 2½in strips cut across the width of the fabric
- 1yd (1m) of fabric for borders and background
- Spare jelly roll strips used for binding

Sorting the Strips
- Choose two light strips and two dark strips for the four-patch blocks.
- Choose four strips for the binding.
- Choose twelve strips for the strip rows.

Cutting Instructions
- Cut five 3in strips across the width of the fabric for the border.
- Cut four 4½in strips across the width of the fabric and sub-cut each into eight 4½in squares.
- Cut across the diagonal of each square to make sixty-four background triangles.

Making the Four-Patch Blocks

1. Take two of the strips you have allocated for the four-patch blocks and sew them together down the long side. Press to the darker side. Repeat with the other two strips allocated for the four-patch blocks.

2. Layer the two strip units right sides together, reversing the lighter and darker fabrics. Align the edges and make sure the centre seams are nesting up against each other. Trim the selvedge and sub-cut into sixteen 2½in segments.

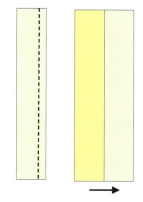

3. Chain piece the segments together. Open out and press the work. You need sixteen four-patch blocks.

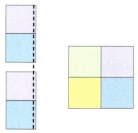

4. With right sides together, sew one of the background triangles to one side of the four-patch block, aligning the edges and making sure the tip of the triangle is pointing to the centre seam – this ensures that the triangle is centred. Open and press towards the four-patch block. Repeat on the other side and press towards the four-patch block.

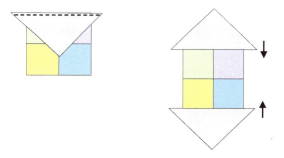

5. Repeat on the two other sides, pressing the seams towards the four-patch block.

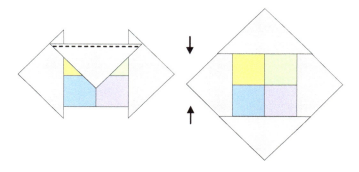

Sherbert Lemon Quilt 17

6. Using a quilting ruler, trim the block to measure 6½in square, trying to ensure the four-patch block is centred. Repeat with the other four-patch blocks to make a total of sixteen 6½in four-patch blocks on point. If you wish you can carefully trim any excess fabric on the reverse to reduce bulk.

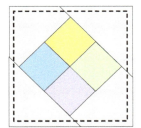

Sewing the Strip Rows

7. Take the strips allocated for the strip rows and sew four of the strips together along the length. Repeat twice more until all twelve strips are used. To prevent any bowing in your strip units it is best to sew two strips together and press before adding the third and then press before adding the fourth. Chain piece whenever you can for speed (see page 113).

8. Press the strip units and trim the selvedge. Cut each strip unit into six 6½in segments. You will have a total of eighteen segments each measuring 6½in x 8½in.

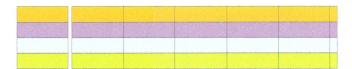

Assembling the Quilt Top

9. Lay out your segments into three rows of six segments each. Rotate some to make sure you don't have the same fabrics too near each other.

10. Place the on point four-patch blocks between the rows of segments. When you are happy with the layout, sew the segments together to form rows and then sew the on point four-patch blocks together to form rows. Press the work.

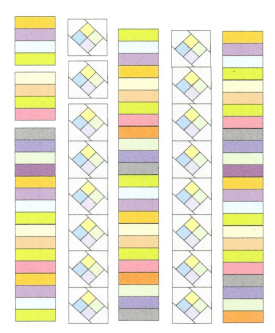

11. With right sides together, sew the rows together pinning at every seam intersection to ensure a perfect match. Press the work.

Adding the Borders and Finishing

12. Join the five 3in border strips into one continuous length and, referring to the instructions on page 117, add the borders to the quilt.

13. Your quilt top is now complete. Quilt as desired (see page 118 for advice) and bind to finish (page 118). If you want to have a scrappy binding cut each of the binding strips into four rectangles 2½in x 10½in and re-sew into a continuous length, alternating fabrics.

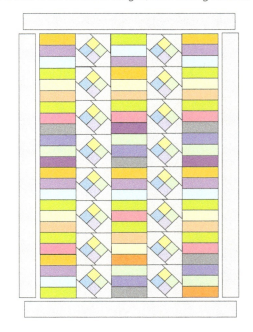

Sherbert Lemon is such a bright and cheerful little cot quilt and the four-patch blocks set on point create a really interesting design. The quilt was pieced by the authors and longarm quilted by The Quilt Room.

Sherbert Lemon Quilt

Chapter Two
Speedwell and Bubblegum

For the two quilts in this chapter we used a jelly roll by Tula Pink from Moda, which had a lovely assortment of aqua, pink and brown and we were really pleased with the effect. Both of these designs are set on point and are good examples of how simple patterns can look so different when turned at an angle.

The setting triangles in Speedwell are made from the strip units and make the quilt look far more complex than it really is – we're all for that! The corner triangles need a bit of juggling to find a pleasing design as they don't follow the pattern exactly.

Bubblegum is a simple framed nine-patch block set on point. Alternating it with a 'not quite plain' block creates very little extra work but turns a simple block into a stunning quilt.

For the Bubblegum Quilt choose your twelve favourite strips to be the outer strips for the nine-patch blocks as they decide the look of the quilt. This quilt displays the nine-patch blocks on point most attractively.

Speedwell Quilt

VITAL STATISTICS
Quilt size: 42in x 49in
Block size: 5½in square
Number of blocks: 32
Setting: on point plus 5in border

REQUIREMENTS
- Half a jelly roll OR twenty 2½in strips cut across the width of the fabric
- Additional 4in (10cm) of colour C (ours was aqua)
- 28in (70cm) of fabric for border
- 16in (40cm) of fabric for binding

Sorting the Strips
- Choose seven strips to be colour A (ours were coffee).
- Choose seven strips to be colour B (ours were pink).
- Choose six strips to be colour C (ours were aqua).
- The seventh strip of colour C (aqua) is cut from the extra 4in (10cm).

Cutting Instructions
Border fabric:
- Cut four 5½in strips across the width of the fabric.

Binding fabric:
- Cut five 2½in wide strips across the width of fabric.

Making the Blocks

1. Take one strip each of colour A, colour B and colour C and sew together along their length, ensuring that colour A is on the top, colour B is in the middle and colour C is on the bottom. Check that the width of your strip is 6½in. If not, adjust the seam allowance. Press seams in one direction.

2. Make seven strip units in total, ensuring that colour A is always on the top of the strip unit, colour B is in the middle and colour C is on the bottom.

3. Take three strip units, trim the selvedge and cut each strip unit into six 6½in squares. Cut each square diagonally *from bottom right to top left*. Put the two different triangles from Cut A in separate piles, keeping them as though they were still in a square. You need to keep the left-hand one on the left and the right-hand one on the right. You will have eighteen triangles in each pile.

Cut A

4. Take three of the strip units, trim the selvedge and cut each strip unit into six 6½in squares. Cut each square diagonally *from bottom left to top right*. Put the two different triangles from Cut B in separate piles, keeping them as though they were still in a square. Keep the left-hand one on the left and the right-hand one on the right. You will have eighteen triangles in each pile. Don't mix them up with the triangles from Cut A.

Cut B

The corner triangles in this quilt do need a little juggling to achieve a pleasing design as they don't follow the pattern exactly. It may take a little time but it will be worth it. The quilt was pieced by the authors and longarm quilted by The Quilt Room.

Speedwell Quilt

5 Take the seventh strip unit, trim the selvedge and cut five 6½in squares. Cut two squares diagonally as in Cut A and three squares diagonally as in Cut B.

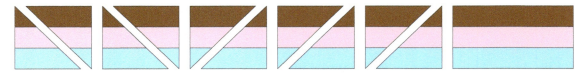

6 Put these triangles on to the correct piles. You should now have twenty triangles in the two Cut A piles and twenty-one triangles in the two Cut B piles, laid out as shown in the diagram below.

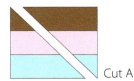 Cut A 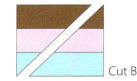 Cut B

7 Now take a right-hand triangle of Cut A and sew it to a left-hand triangle of Cut B, ensuring that the strips of colour A are on the outside of the block as shown. Make twenty blocks.

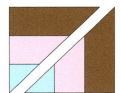  x20

8 Take a left-hand triangle of Cut A and sew it to a right-hand triangle of Cut B, ensuring that the strips of colour C are on the outside of the block as shown. Make twelve blocks. Do not sew the remaining triangles as these are the setting and corner triangles.

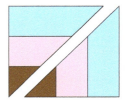 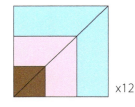 x12

Assembling the Blocks

9 Referring to the quilt diagram opposite, lay out the blocks with a setting triangle at each end. Make sure the setting triangles are facing the correct way. You will find that when placing your setting triangles, your triangles from Cut A (left) fit the left-hand side of the quilt, and Cut B (right) fit the right-hand side of the quilt. The top and bottom of the quilt require alternating triangles. The four remaining ones fit the corners.

10 Once you are happy with the layout, sew the blocks into rows with a setting triangle at each end. Always press the seams towards the setting triangles and the block with colour C on the outside as shown in the quilt diagram and the seams will then nest together nicely.

11 The setting triangles are slightly larger, so when sewing these make sure the bottom of the triangle is aligned with the block, as shown in the diagram below.

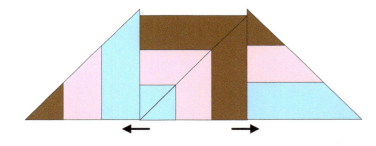

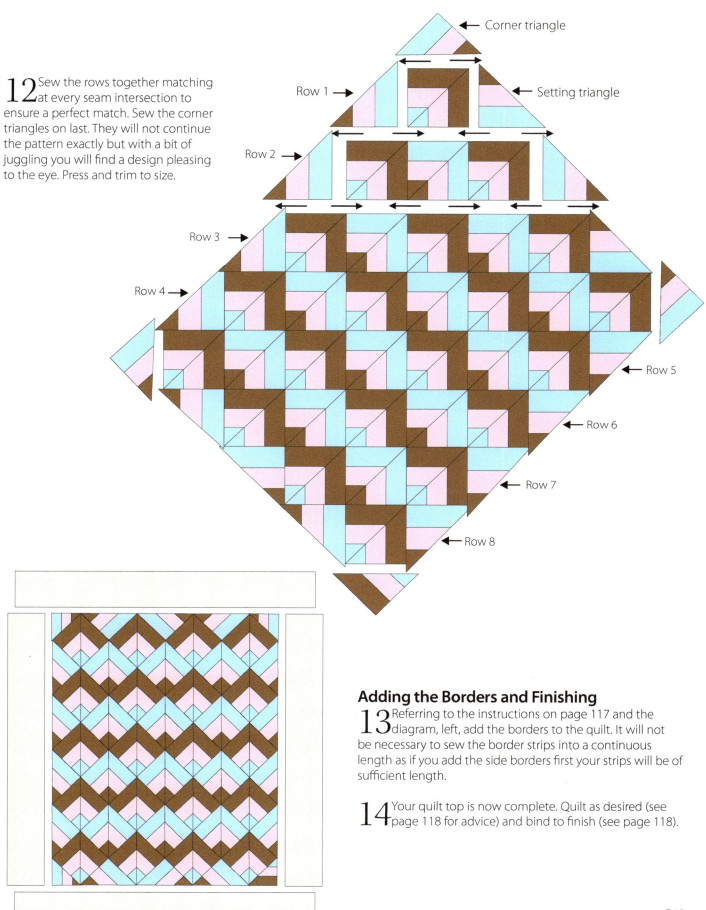

12 Sew the rows together matching at every seam intersection to ensure a perfect match. Sew the corner triangles on last. They will not continue the pattern exactly but with a bit of juggling you will find a design pleasing to the eye. Press and trim to size.

Adding the Borders and Finishing

13 Referring to the instructions on page 117 and the diagram, left, add the borders to the quilt. It will not be necessary to sew the border strips into a continuous length as if you add the side borders first your strips will be of sufficient length.

14 Your quilt top is now complete. Quilt as desired (see page 118 for advice) and bind to finish (see page 118).

Speedwell Quilt 25

Bubblegum Quilt

VITAL STATISTICS
Quilt size: 42in x 56in
Block size: 10in square
Number of blocks: 18
Setting: on point

REQUIREMENTS
- Half a jelly roll OR twenty 2½in wide strips cut across the width of the fabric
- 1½yd (1.4m) of fabric for background
- 16in (40cm) of fabric for binding

Sorting the Strips
- Choose twelve jelly roll strips to be the outer strips for the nine-patch blocks. Each of these twelve strips will be the outer strip of one nine-patch block.
- The six little centre squares in the alternate blocks are also cut from six of these strips.
- The remaining eight jelly roll strips will make the nine-patch blocks.

Cutting Instructions
Jelly roll strips:
- From the twelve jelly roll strips to be used as the outer strips for the nine-patch blocks cut each into two rectangles 2½in x 6½in and two rectangles 2½in x 10½in. Keep the rectangles from each strip together.
- Cut six 2½in squares from the offcuts for the centre of the alternate blocks.

Background fabric:
- Cut four 4½in wide strips across the width of the fabric. Sub-cut three into 4½in x 10½in rectangles. You get four to each strip. You need twelve in total.
  Sub-cut one strip into 2½in x 4½in rectangles. You need twelve.
- Cut two 16in wide strips across the width of the fabric. Sub-cut one of these strips into two 16in squares. Sub-cut the other into one 16in square and then trim the remainder of the strip to measure 8½in wide and cut two 8½in squares.
- Take the three 16in squares and cut across both diagonals of each square to create ten setting triangles. You will have two spare.

- Take the two 8½in squares and cut across one diagonal of each square to create four corner triangles. Cutting the setting and corner triangles in this way ensures that there are no bias edges on the outside of your quilt.

16in square   8½in square

Binding:
- Cut five 2½in wide strips across the width of the fabric.

Making the Nine-Patch Blocks
1. Pair up the eight jelly roll strips allocated for the nine-patch blocks. Take one pair of strips and cut each into three lengths of 14in.

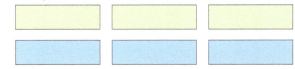

2. From your six lengths sew two strip segments as shown below. Press to the darker fabric. Cut each strip unit into five 2½in segments.

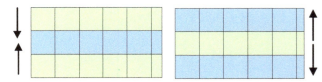

3. Sew together three nine-patch blocks. The extra strip is spare. Press the work. Repeat with the three other pairs of strips to make twelve nine-patch blocks.

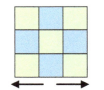

Note: if you want a scrappier effect for the nine-patch blocks you can cut your eight strips into 14in lengths and then join them together randomly.

Adding the Outer Strips

4 Sew a 2½in x 6½in strip of the same fabric to the top and bottom of the nine-patch block. Press the seams away from the nine-patch block.

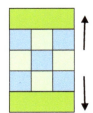

5 Sew a 2½in x 10½in strip of the same fabric to either side of this unit. Press the seams away from the nine-patch block. Repeat and add the outer strips to all twelve nine-patch blocks.

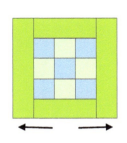

Making the Alternate Blocks

6 Sew a background 2½in x 4½in rectangle to either side of a 2½in square allocated for the centres of the alternate block. Press towards the background.

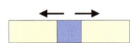

7 Sew a background 4½in x 10½in rectangle to the top and bottom of this unit. Press to the background. Repeat to create six alternate blocks.

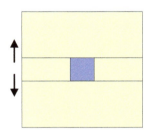

Assembling the Quilt

8 To set the blocks on point refer to the quilt diagram (above right), laying out the blocks as shown. When you are happy with the layout sew a setting triangle to each side of a nine-patch block to create Row 1. The setting triangles have been cut slightly larger to make the blocks 'float', so when sewing the setting triangles make sure the bottom of the triangle is aligned with the block. Press as shown below.

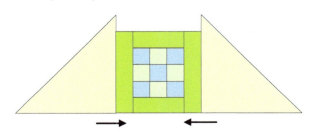

9 Continue to sew the blocks together to form rows with setting triangles at each end. Always press towards the nine-patch block as this will ensure your seams are going in different directions when sewing the rows together.

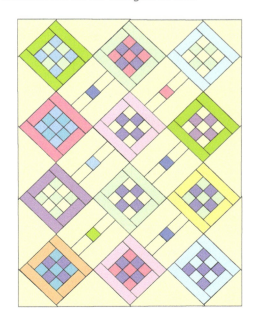

10 Sew the rows together, pinning at every intersection, and sew the corner triangles on last.

11 Your quilt top is now complete. Quilt as desired (see page 118 for advice) and bind to finish (page 118).

Setting a nine-patch block on point makes the squares look like diamonds, which always looks far more complex – people will think you are very clever! The quilt was pieced by the authors and longarm quilted by The Quilt Room.

Bubblegum Quilt

Chapter Three
Square Dance and Strip the Willow

These two quilts use the same jelly roll but look quite different, mostly due to the very different scale used in the blocks. As much as we love intricate patterns we would never consider cutting squares as small as ¾in. Square Dance, however, does indeed have ¾in sashing squares but we definitely didn't cut them like that – the ever-versatile nine-patch block helped out again. Yes, there are lots of blocks to sew together but it goes together easily with a bit of organized chain piecing – and lots of pins!

The second quilt, Strip the Willow, is a bright and cheerful rail fence variation which makes good use of every inch of fabric. If you want a quilt in a hurry then look no further.

Strip the Willow is a quilt that you really can sit down and accomplish in an afternoon – yes really! Just chain piece those strips together and away you go.

Square Dance Quilt

VITAL STATISTICS
Quilt size: 39in x 51in
Block size: 2¾in square
Number of blocks: 238
Setting: on point

REQUIREMENTS
- Half a jelly roll OR twenty 2½in strips cut across the width of the fabric
- 1¼yd (1.10m) of fabric for background
- 16in (40cm) of fabric for setting triangles (this could be the same as the background)
- 16in (40cm) of fabric for binding

Sorting the Strips
- Choose four strips for the sashing squares – we used one colourway (red).
- The remaining sixteen strips are used in the nine-patch blocks.

Cutting Instructions
Background fabric:
- Cut sixteen 2½in wide strips across the width of the fabric.

Setting triangles:
- Cut two 5¾in wide strips across the width of the fabric. Sub-cut one of these strips into seven 5¾in squares. Sub-cut the second strip into four 5¾in squares to make a total of eleven 5¾in squares.
Trim the remainder of the second strip to 3½in wide and cut two 3½in squares.
- Take the eleven 5¾in squares and cut across both diagonals of each square to create forty-two setting triangles. You will have two spare.
- Take the two 3½in squares and cut across one diagonal of each square to create four corner triangles.

5¾in square   3½in square

Cutting the setting and corner triangles this way ensures there are no bias edges on the outside of your quilt.

Binding:
- Cut five 2½in wide strips across the width of the fabric.

Making the Blocks

1. Sew two jelly roll strips allocated for the nine-patch blocks to either side of a background strip. Press the seams in one direction. Trim the selvedge and sub-cut into sixteen 2½in segments.

2. Repeat to make a total of eight of these strip units and cut each into sixteen segments. You will have 128 segments. This is unit A.

Unit A

3. Take two background strips and sew to either side of one of the jelly roll strips allocated for the sashing squares. Press the seams in one direction. Sub-cut into sixteen 2½in segments. Repeat to make a total of four of these strip units and cut each into sixteen 2½in segments. You will have sixty-four of these segments. This is unit B.

Unit B

4. With right sides together, sew two unit As to either side of a unit B to make a nine-patch block, pinning at every seam intersection to ensure a perfect match. Do not concern yourself with co-ordinating strips as you will be cutting them apart next! Press seams in one direction. Repeat to make sixty blocks. You will have four spare.

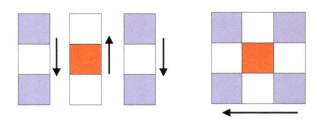

We love the versatile nine-patch block and here again it comes to our aid. The nine-patch blocks are dividing into four, creating this intricate looking pattern. The quilt was pieced by the authors and longarm quilted by The Quilt Room.

Square Dance Quilt

5. Cut each nine-patch block accurately through the centre in each direction to make four smaller blocks. The centre segment is 2in wide so the best way of cutting accurately is to line up the 1in mark on the ruler on the seam line. Make a total of 238 smaller blocks. Two will be spare.

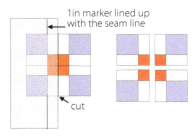

Assembling the Quilt

6. Sew the blocks together to form rows before laying out the quilt. Try not to put the same fabric next to each other but pay more attention to which seams are going in opposite directions in order that they nest together nicely. Chain piece for speed by sewing blocks into pairs and then pairs into sets of four, and so on. As can be seen from the quilt diagram below, there are four rows of nineteen blocks, two rows of seventeen, two rows of fifteen, two rows of thirteen, and so on. Don't press the rows yet.

7. Lay out your rows and sew a setting triangle to both ends of each row. Don't sew the corner triangles on yet. Press seams of alternate rows in opposite directions. The setting triangles are cut slightly larger to make the blocks 'float', so when sewing the triangles on rows 1–9 make sure the bottom of the triangle is aligned with the *bottom* of the block. On rows 14–22 align the bottom of the setting triangles with the *top* of the block. On rows 10–13 refer to the quilt diagram for how the triangles should be aligned.

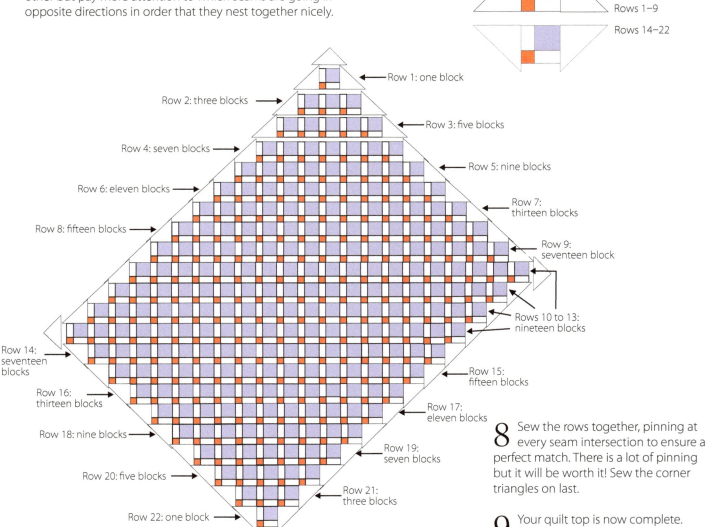

8. Sew the rows together, pinning at every seam intersection to ensure a perfect match. There is a lot of pinning but it will be worth it! Sew the corner triangles on last.

9. Your quilt top is now complete. Quilt as desired (see page 118 for advice) and bind to finish (page 118).

Strip the Willow Quilt

VITAL STATISTICS
Quilt size: 43in x 49in
Block size: 6½in square
Number of blocks: 30
Setting: 5 x 6 plus 1in inner border and 4in outer border

REQUIREMENTS
- Half jelly roll OR twenty 2½in wide strips cut across the width of the fabric
- 10in (25cm) of fabric for inner border
- 20in (50cm) of fabric for outer border
- 16in (40cm) of fabric for binding

Sorting the Strips
- Divide the strips into five sets of four strips each.
- Choose one fabric from each set to be the centre strip, one fabric to form the two narrow strips and two to be the outer strips.

Cutting Instructions
Jelly roll strips:
- Trim the five centre strips lengthways to 2in wide.
- Take the five strips allocated to form the narrow strips and cut each in half lengthways to create ten strips 1¼in wide. Leave the remaining ten strips uncut.

Inner border fabric:
- Cut four 1½in wide strips across the width of the fabric.

Outer border fabric:
- Cut four 4½in wide strips across the width of the fabric.

Binding:
- Cut five 2½in wide strips across the width of the fabric.

This variation of a Rail Fence block creates an interesting pattern. The quilt was pieced by the authors and longarm quilted by The Quilt Room.

Sewing the Strips

1. Make a strip unit by sewing the strips together as shown, starting with sewing a narrow 1¼in wide strip to a 2½in wide strip. Add the 2in wide strip, the second narrow strip and then the other 2½in wide strip. Repeat to make five strip units in total.

   You don't want any bowing in your strip units so it is best to sew two strips together and press before adding the third. Press before adding the fourth and press again before adding the fifth. Chain sew whenever you can. Press all seams in one direction and trim the selvedge. Take care not to trim excessively as you need 42in to make the blocks.

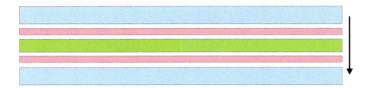

2. Cut each of the five strip units into six 7in segments and then trim each segment to measure 7in square. You will have a total of thirty segments. When trimming each segment to measure 7in square, trim a little from each side to balance the block.

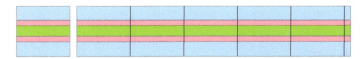

Assembling the Blocks

3. Lay out the blocks as shown, alternating every alternate block 90 degrees and when you are happy with the arrangement sew the blocks into rows and then sew the rows together, pinning at every seam intersection to ensure a perfect match. Press the seams of row 1 to the right and the seams of row 2 to the left and you will find the seams nest together nicely.

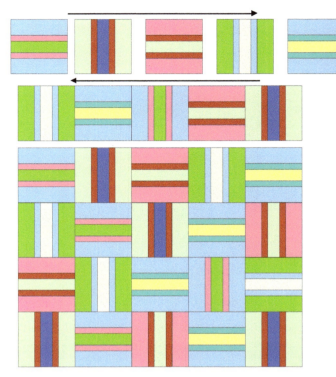

Adding the Borders and Finishing

4. Referring to the instructions on page 117, add the inner borders to the quilt. There is no need to join the strips into a continuous length. Repeat to add the outer borders.

5. Your quilt top is now complete. Quilt as desired (see page 118 for advice) and bind to finish (page 118).

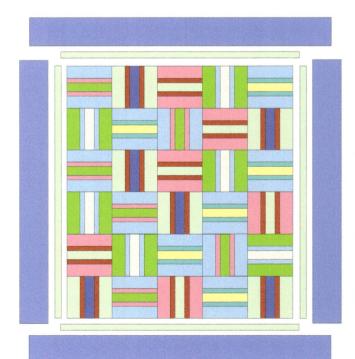

Chapter Four
Sailing Boats and Roller Coaster

We loved this bright, cheery range by American Jane for Moda, which has created two fabulously colourful quilts, although the strips with the 'Happy Campers' on – some without heads and others without legs – did present a challenge!

Sailing Boats uses a traditional boat block, combined with a sixteen-patch block to create a fun quilt. We picked our fabrics for this quilt first, still wondering what we were going to do with our headless little people!

As it turned out the 'problem' strips look fine in Roller Coaster and in fact they add interest – the lack of a head in one strip is made up for by three legs on another! It definitely gets a second glance. The moral here is that if you do have some slightly odd fabrics in your jelly roll, don't panic – just work with them and you will be amazed at the results.

Sailing Boats Quilt

VITAL STATISTICS
Quilt size: 44in x 52in
Block size: 8in square
Number of blocks: 30
Setting: 5 x 6 blocks plus 2in border

REQUIREMENTS
- Half a jelly roll OR twenty 2½in strips cut across the width of the fabric
- 10in (25cm) of fabric for the boats
- 1½yd (1.25m) of fabric for background and border
- 16in (40cm) of fabric for binding
- Omnigrid 96 ruler or other speciality ruler

Sorting the Strips
- Choose sixteen strips for the sixteen-patch block.
- Choose two and a half strips for the sails – these can be five different half strips.

Cutting Instructions
Jelly roll strips:
- Cut nineteen jelly roll strips in half to make thirty-eight rectangles 2½in x 21in. You need thirty-seven half strips so you will have one half strip spare plus one whole strip.

Fabric for boat:
- Cut three 2½in wide strips across the width of the fabric.
- Sub-cut each strip into five rectangles 2½in x 8½in. You need fifteen in total.

Background fabric:
- Cut eighteen 2½in wide strips cut across the width of the fabric.
  Sub-cut three in half to make six rectangles 2½in x 21in. You need five to make the sail units – one is spare.
  Sub-cut two strips into thirty 2½in squares to make the boat unit.
  Sub-cut four strips to make thirty rectangles 2½in x 4½in. You get eight to a strip – so two are spare.
  Sub-cut four strips to make fifteen rectangles 2½in x 8½in. You get four to a strip – so one is spare.
- Set five strips aside for the border.

Binding:
- Cut five 2½in strips across the width of the fabric.

Making the Sail Units

1. Take one jelly roll half strip allocated for the sails and one background half strip and press right sides together ensuring that they are exactly one on top of the other. The pressing will help hold the two strips together.

2. Lay out on a cutting mat and trim the selvedge on the left side. Position the Omnigrid 96 ruler as shown in the diagram, lining up the 2in mark at the bottom edge of the strips, and cut the first triangle. You will notice that the cut out triangle has a flat top. This would just have been a dog ear you needed to cut off – so it is saving you time!

3. Rotate the ruler 180 degrees to the right and cut the next triangle. Continue along the strip. You need twelve sets of triangles from one strip unit.

4. Sew along the diagonals to form twelve half-square triangles. Trim all dog ears and press open with seams pressed towards the darker fabric.

5. With right sides together, sew two units together and press. Repeat with another set of units, pressing in the opposite direction. Sew the two units together pinning at the seam intersection to ensure a perfect match. Repeat to make three sets of sails and press.

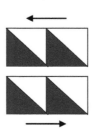

This bright and cheerful quilt is straightforward to make and would be a wonderful quilt for a boy's bed. The quilt was pieced by the authors and longarm quilted by The Quilt Room.

Sailing Boats Quilt

6. Repeat with the other four jelly roll half strips allocated for the sails and four background half strips to make in total fifteen sets of sails.

Making the Boat Units

7. Take one 2½in background square and lay it right sides together on a 2½in x 8½in boat rectangle. Sew across the diagonal. If it helps, draw the diagonal line in first or make a fold to mark your stitching line. Flip the square over. Press towards the background fabric. Trim excess background fabric but do not trim the boat fabric. Although this creates a little more bulk it helps to keep the work in shape.

8. Sew another background square to the other side of the boat rectangle. Press and trim the excess background fabric. Repeat to make fifteen boats.

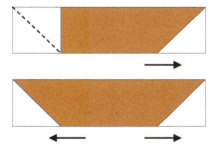

Assembling the Sailing Boat Blocks

9. Take one set of sails and sew a 2½in x 4½in background rectangle to either side. Press to the background fabric.

10. Sew this unit to a boat unit, pinning at every seam intersection to ensure a perfect match. Press towards the boat unit.

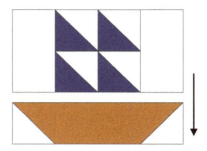

11. Sew a 2½in x 8½in background rectangle to the bottom of the unit. Press to the background fabric. Repeat to make fifteen sailing boat blocks.

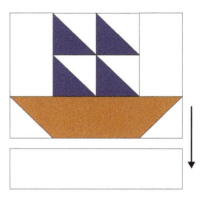

Making the Sixteen-Patch Blocks

12. Select four jelly roll half strips and sew together. Press seams in one direction. Repeat to form eight strip units in total.

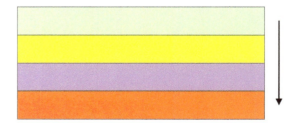

13. Sub-cut each strip unit into eight 2½in segments. You need sixty in total – four are spare.

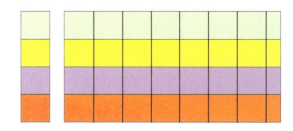

14 Select four segments and sew together, pinning at every seam intersection to ensure a perfect match. Try to avoid having two squares of the same colour next to each other if possible.

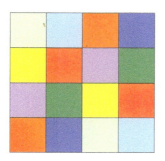

Assembling the Quilt

15 Lay out your blocks into rows, alternating the sailing boat block with the sixteen-patch block. Sew the blocks into rows and then sew the rows together. Pin at every seam intersection to ensure you have matching seams. Press row 1 with the seams to the left and press the seams of row 2 with the seams to the right – this way they will nest together nicely.

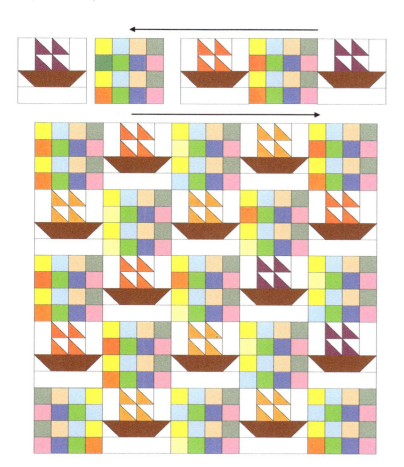

Adding the Borders and Finishing

16 Referring to the instructions on page 117, add the borders to the quilt. If you add the top and bottom borders first, you only need to join three border strips into a continuous length so you will only have joins in your side borders.

17 Your quilt top is now complete. Quilt as desired (see page 118 for advice) and bind to finish (see page 118).

Sailing Boats Quilt

Roller Coaster Quilt

VITAL STATISTICS
Quilt size: 37in x 48in
Block size: 5½in square
Number of blocks: 48
Setting: 6 x 8 blocks plus a 2in border

REQUIREMENTS
- Half a jelly roll OR twenty 2½in strips cut across the width of the fabric – you need twelve dark and twelve light strips so your additional fabric is dependent on what your jelly roll contains
- 12in (30cm) of additional light or dark fabric
- 16in (40cm) of fabric for border
- 16in (40cm) of fabric for binding

Sorting the Strips
- Divide your jelly roll, plus the four strips made from the additional fabric, into twelve light strips and twelve dark strips.

Cutting Instructions
Additional fabric:
- Cut four 2½in wide strips across the width of the fabric.

Border fabric:
- Cut five 2½in wide strips across the width of the fabric.

Binding fabric:
- Cut five 2½in wide strips across the width of fabric.

Making the Blocks

1. Sew three assorted dark strips together along their length. Check that the width of your strip is 6½in. If not then adjust your seam allowance. Press the seams in one direction and trim the selvedge.

2. Cut the strip units into six 6½in squares – you will get six per strip unit. Repeat to make another three strip units. You need twenty-four 6½in dark squares in total.

3. Repeat the above with the twelve light strips, making a total of twenty-four 6½in light squares. Press seams in the opposite direction.

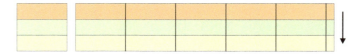

This quilt is certainly eyecatching with all those fabulous reds, yellows and blues and is a super-quick quilt to make. The quilt was pieced by the authors and longarm quilted by The Quilt Room.

Roller Coaster Quilt

4. Draw a diagonal line on the reverse side of twelve light squares, with the diagonal line going from *bottom left to top right*.

5. Draw a diagonal line on the reverse side of the other twelve light squares, this time with the diagonal line going from *bottom right to top left*.

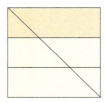

6. Lay a light square marked with the line from bottom left to top right on top of a dark square aligning the edges and ensuring the seams are butting up to each other. Keep the squares the same way up and make sure all the seams are horizontal.

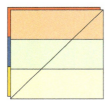

7. Stitch either side of the drawn line with a scant ¼in seam allowance. Repeat with all the twelve light squares with the line drawn from bottom left to top right. Press the stitches to set and then cut along the drawn diagonal line. Trim dog ears and press open towards the dark fabric.

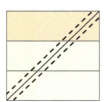

8. These will be Block A (shown in diagram below). You need twenty-four. Notice that there are two different blocks but the important thing is the placement of the light and dark triangles and the direction of the seams.

Block A

9. Repeat steps 6–8 using the light squares, which are marked with the diagonal line going from bottom right to top left.

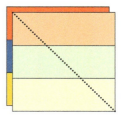

10. These will be Block B. You need twenty-four. Notice again that there are two different blocks but the important thing is the placement of the dark triangles and the direction of the seams.

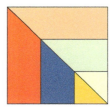

Block B

Assembling the Blocks

11. Before laying out your blocks, you might find it useful to label them A and B or distinguish them by a different coloured pin or coloured sticker.

12. Referring to the diagram, right, sew the blocks together to create rows 1 and 2, making sure you are positioning blocks A and B correctly to create the design. Sew the blocks to make rows 3 and 4, which are identical to rows 1 and 2.

13. Press the seams in rows 1 and 3 to the left and the seams in rows 2 and 4 to the right to ensure the seams nest together when sewing the rows together.

14. Sew the rows together, pinning at every seam intersection to ensure a perfect match. This is the top half of your quilt top.

15. Repeat the four rows to make the bottom half of your quilt top. Rotate this bottom half 180 degrees and sew the top and bottom halves together, pinning at every intersection to ensure a perfect match.

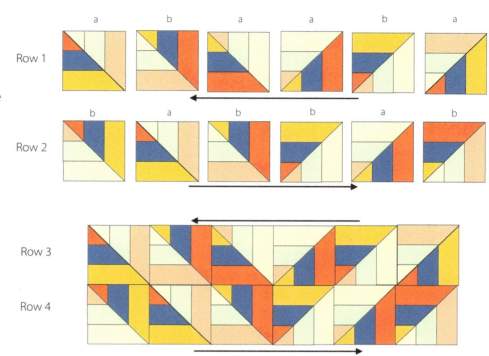

Adding the Borders and Finishing

16. Referring to the instructions on page 117 and the diagram, left, add the borders to the quilt. If you add the top and bottom borders first, you only need to join three border strips into a continuous length so you will only have joins in your side borders.

17. Your quilt top is now complete. Quilt as desired (see page 118 for advice) and bind to finish (see page 118).

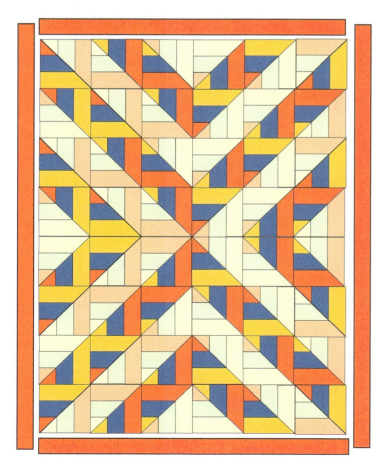

Roller Coaster Quilt

Chapter Five
Stardust and Playmates

We just loved the wonderful assortment of greys and reds in this jelly roll, especially as the reds were a combination of bright reds drifting into subtle faded reds. The range is called Rouenneries and is by French General from Moda.

For our first quilt, Stardust, we pulled out all the brighter reds and the lighter strips, saving the fabulous faded reds and an assortment of lovely greys and lights for the Playmates quilt. The two quilts are similar in style and complement each other beautifully, although are made using quite different techniques. The Stardust quilt needs just a little more attention when choosing the strips so we have described how to make that quilt first.

Stardust Quilt

VITAL STATISTICS
Quilt size: 41in x 41in
Block size: 8in square
Number of blocks: 9
Setting: 3 x 3 blocks plus pieced 4in sashing and 4½in border

REQUIREMENTS
- Half a jelly roll OR twenty 2½in strips cut across the width of the fabric
- 35in (90cm) of fabric for centre squares and the border
- Spare jelly roll strips and excess border fabric to be used for the binding
- Omnigrid 96 ruler or other speciality ruler

Sorting the Strips
- Choose five dark strips for the star points.
- Choose eight light strips for the background.
- Choose four medium/dark strips for the four-patch blocks.
- Use the remaining three strips in the binding.

Cutting Instructions
Jelly roll strips:
- Take two of the light jelly roll strips and cut each into sixteen 2½in squares. Take a third light strip and cut four 2½in squares. You need thirty-six 2½in squares in total.
- Cut the balance of the third strip (approximately 32in) into three rectangles 2½in x 10½in and set aside for the binding. Leave the remaining light strips uncut.
- Take the three strips allocated for the binding and cut each into four rectangles 2½in x 10½in.

Centre squares and border fabric:
- Cut four 5in strips across the width of the fabric and set these aside for the border.
- Cut two 4½in strips across the width of the fabric. Sub-cut these into twelve 4½in squares.
- Cut one 2½in wide strip across the width of the fabric and sub-cut into four rectangles 2½in x 10½in and set these aside for the binding.

Making the Four-Patch Units
1. Take two of the strips allocated for the four-patch units and lay them right sides together, as shown in the diagram below. Sew down the long side, open and press towards the darker fabric. Repeat with the other two strips.

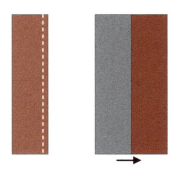

2. With right sides together, lay one strip unit on top of another, with the lighter strip on the top of one unit and on the bottom of the other, ensuring that the centre seams are in alignment. Trim the selvedge and sub-cut these strips into thirteen 2½in wide segments.

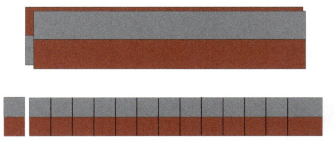

3. Keep the pairs together and sew down the long side as shown. Press open to form thirteen four-patch units. These are unit A.

Unit A

This striking quilt uses the brighter reds from the jelly roll, coupled with the lighter strips. Four-patch blocks and half-square triangles create the star pattern. Our quilt doesn't have the scrappy binding given in the instructions and if you prefer this you will need an extra 16in, cut into five 2½in wide strips across the fabric width. The quilt was pieced by the authors and longarm quilted by The Quilt Room.

Stardust Quilt

Making the Half-Square Triangle Units

4 Press one dark and one light strip right sides together, ensuring that they are exactly one on top of the other. The pressing will help to hold the two strips together.

5 Lay the strips out on a cutting mat and trim the selvedge on the left side.

6 Position the Omnigrid 96 ruler as shown in the diagram below, lining up the 2in mark at the bottom edge of the strips, and cut the first triangle. You will notice that the cut out triangle has a flat top. This would just have been a dog ear you needed to cut off – so it is saving you time!

2in line

7 Rotate the ruler 180 degrees to the right as shown and cut the next triangle. Continue along the strip. You need twenty-four pairs of triangles from one strip.

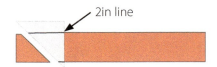

2in line

8 Sew along the diagonals to form twenty-four half-square triangle units. Trim all dog ears and press open, pressing twelve to the dark fabric and twelve to the light.

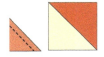

9 With right sides together sew a unit pressed to the dark to a unit pressed to the light, ensuring the points are matching. Repeat to make twelve units. These are unit B.

10 Repeat with all five dark and light strips. You need sixty of unit B in total.

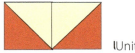

(Unit B

Assembling the Star Block

11 Sew a 2½in light square to either side of a unit B as shown. Press in the directions shown below.

12 Sew a unit B to either side of a unit A and press as shown below. Sew a 2½in light square to either side of a unit B and press.

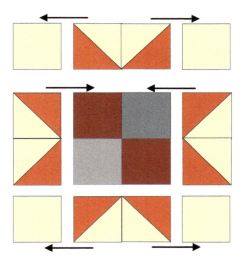

13 Now sew the rows together, pinning at every seam intersection to ensure a perfect match. Repeat to make nine of the star blocks.

Assembling a Sashing Unit

14 Sew a unit B to either side of a 4½in square and press. Repeat to make twelve of the sashing units.

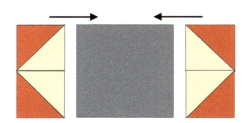

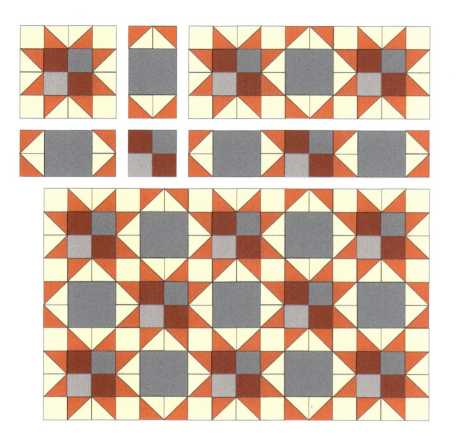

Assembling the Quilt Top

15 Referring to the diagram, left, lay out the star blocks with the sashing units and the remaining unit As. When you are happy with the arrangement, sew the units together, pinning at every seam intersection. Sew the rows together, again pinning at every seam intersection to ensure that the seams match nicely.

Adding the Borders and Finishing

16 Add your borders referring to the diagram, left, and the instructions on page 117. Your quilt top is now complete, so quilt as desired (see page 118 for advice).

17 Join all the 2½in x 10½in rectangles into a continuous length alternating the different fabrics to make the binding and bind to finish (see page 118).

Playmates Quilt

VITAL STATISTICS
Quilt size: 43in x 58in
Block size: 5½in square
Number of blocks: 35
Setting: 5 x 7 blocks plus 2in sashing strips and a 2in border

REQUIREMENTS
- Half a jelly roll OR twenty 2½in strips cut across the width of the fabric
- 1½yd (1.20m) of fabric for sashing and border
- 16in (40cm) of fabric for binding

Sorting the Strips
- Choose two strips for the sashing squares.
- The remaining eighteen will be used in the blocks.

Cutting Instructions
Jelly roll strips:
- Take the two strips allocated for the sashing squares and cut each into sixteen 2½in squares.
- Take the eighteen strips allocated for the blocks and cut one 2½in square from each.
- You need forty-eight 2½in sashing squares in total – you will have two spare.

Sashing and border fabric:
- Cut seventeen strips 2½in wide across the width of the fabric.
- Set five strips aside for the borders.
- Sub-cut each of the remaining twelve strips into seven rectangles 2½in x 5½in each for the sashing. You need eighty-two in total – you will have two spare.

Binding fabric:
- Cut five strips 2½in wide across the width of the fabric.

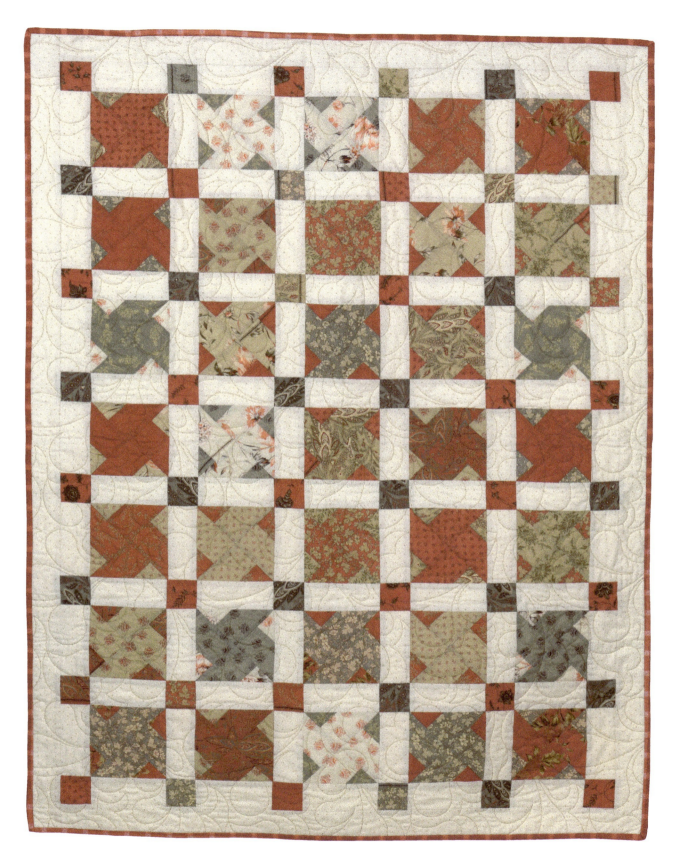

This little quilt has a timeless quality, featuring as it does some lovely faded reds and an assortment of pretty greys and lights. The pieced sashing strips link the blocks together attractively. The quilt was pieced by the authors and longarmed quilted by The Quilt Room.

Playmates Quilt

Making the Blocks

1. Take the eighteen jelly roll strips to be used for the blocks and pair them into one light strip and one dark strip.

2. Take one pair of light and dark strips and with right sides together, sew down the long side, as shown in the diagram below. Press to the darker fabric.

3. With the dark fabric on the top, cut each strip unit into eight 4½in squares.

4. Cut across the diagonal of all eight squares from bottom right to top left. It is important that all cuts are made in the same direction.

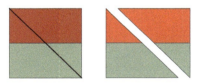

5. Make a pile of the eight left-hand cut triangles and a separate pile of the eight right-hand cut triangles.

6. Take four of the left-hand cut and lay out as shown in the diagram below. Sew the top two together and the bottom two together. Press as shown – the outside edges of the square are on the bias so press gently and do not use steam. Sew the top to the bottom, pinning at the seam intersection to ensure a perfect match. Press the work.

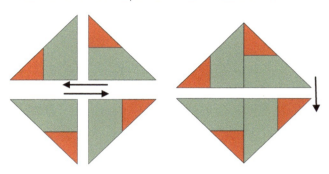

7. Repeat this process with the other four triangles from the left-hand cut. You now have two squares with the light fabric forming a cross in the centre of the square.

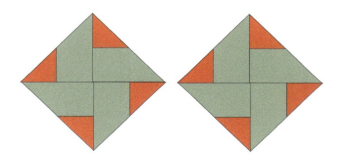

8. Repeat with the triangles from the right-hand cut to make two squares with the dark fabric forming a cross in the centre of the square.

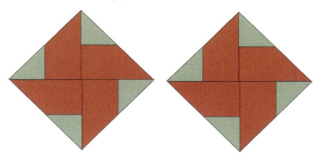

9. Measure your squares – they should be 5½in square. If necessary you can trim them at this stage.

10. Repeat steps 1–9 with the remaining eight pairs of jelly roll strips, always making sure that the dark fabric is on the top and you always cut across the diagonal in the same direction. You need thirty-five squares – one will be spare.

Assembling the Quilt

11. Create the first row by sewing a 2½in square to the left-hand side of five 2½in x 5½in sashing rectangles. Sew a 2½in square to the right-hand side of the last rectangle. Make eight of these rows.

12. Create the second row by sewing a 2½in x 5½in sashing rectangle to the left-hand side of five blocks, alternating the light and dark blocks. Sew a sashing rectangle to the right-hand side of the last block. Make seven of these rows.

13. Referring to the diagram below, sew the rows of the quilt together, pinning at every seam intersection to ensure a perfect match.

Adding the Borders and Finishing

14. Add your borders referring to the diagram below and the instructions on page 117.

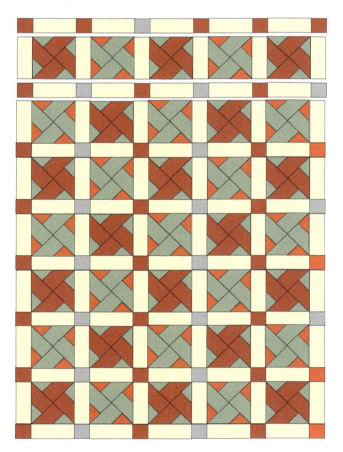

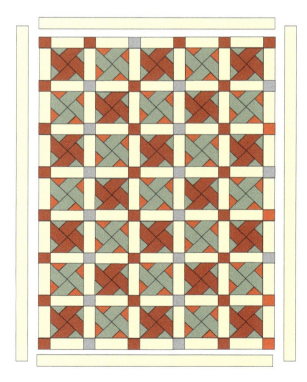

15. Your quilt top is now complete. Quilt as desired (see page 118 for advice) and bind to finish (page 118).

Chapter Six
Snapdragon and Fairy Steps

This Three Sisters Martinique range from Moda has such a lovely mix of aqua, pink, red and grey that we couldn't resist using it for these two beautiful quilts. Snapdragon has a simple nine-patch block alternating with a snowball block and you have enough jelly roll strips left over to create a pieced border. You just need to choose four strips of a similar colour for the flip-over corners and apart from that anything goes.

When you are working with fabrics you just love it is sometimes nice to have a design that just shows off the fabric without too much else and Fairy Steps does just that. We chose a subtle aqua sashing fabric but you could use a bolder colour to create a very different look.

Snapdragon Quilt

VITAL STATISTICS
Quilt size: 38in x 50in
Block size: 6in square
Number of blocks: 35
Setting: 5 x 6 blocks plus 2in inner border and 2in stripe border

REQUIREMENTS
- Half a jelly roll OR twenty 2½in wide strips cut across the width of the fabric
- 1yd (1m) of fabric for background and inner border
- 16in (40cm) of fabric for binding

Sorting the Strips
- Select twelve strips for the nine-patch blocks.
- Select four strips for the snowball block corners.
- Select four strips for the pieced border.

Cutting Instructions
Background fabric:
- Cut three 6½in wide strips across the width of the fabric.
- Sub-cut each of these strips into six 6½in squares. You need seventeen in total for the snowball blocks so one will be spare.
- Cut five 2½in wide strips across the width of the fabric. Set one aside for the outer stripe border and the remaining four for the inner border.

Corners for snowball blocks:
- Take the four jelly roll strips allocated for the snowball block corners and cut each strip into seventeen 2½in squares. This will make sixty-eight 2½in squares. Don't worry if you can't cut as many as seventeen squares from each strip as there are some spare strips when making the nine-patch blocks and the squares from these can be utilized.

Binding:
- Cut five 2½in wide strips across the width of the fabric.

Making the Nine-Patch Blocks

1. Pair up the twelve jelly roll strips allocated for the nine-patch blocks. Take one pair of strips and cut each into three lengths of 14in.

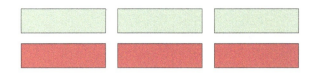

2. From your six lengths sew two strip segments as shown in the diagram below. Press towards the darker fabric. Cut each into five 2½in segments.

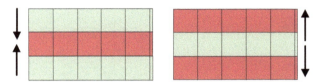

3. Make three nine-patch blocks and set aside the extra strip. Press as shown in the diagram. Repeat with the other sets of strips. You need eighteen nine-patch blocks. From the six extra strips, allocate two strips to add to the stripe borders. The remaining four strips are spare and the squares could be unpicked, if required, to make the number of 2½in squares for the snowball corners up to sixty-eight.

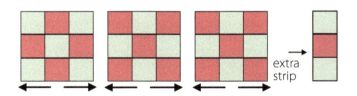

Making the Snowball Blocks

4. Mark a diagonal line from corner to corner on the wrong side of the 2½in square snowball corners.

5. With right sides together, lay a marked square on one corner of a 6½in background square, aligning the outer edges. Sew across the diagonal, using the marked diagonal line as the stitching line. After a while you may find you do not need to draw the line as it is not difficult to judge the sewing line. An alternative is to fold the square and use the fold to guide you. Repeat on the other three corners.

For this lovely quilt we picked out four pinks for the flip-over corners of the snowball blocks. Everything else just fell into place as the fabrics blended so beautifully, with the flashes of red adding just the right amount of sparkle. The quilt was pieced by the authors and longarm quilted by The Quilt Room.

6 Flip the square over and press towards the outside of the block. Trim the excess fabric from the snowball corner but do not trim the background fabric. Although this creates a little more bulk, the background fabric helps keep your patchwork in shape. Repeat with all corners to make seventeen snowball blocks.

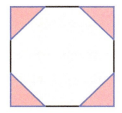

Sewing the Blocks Together

7 Referring to the diagram below, join the blocks together, alternating the snowball blocks with the nine-patch blocks. Pin the point where the snowball triangle joins the nine-patch block to ensure a perfect match.

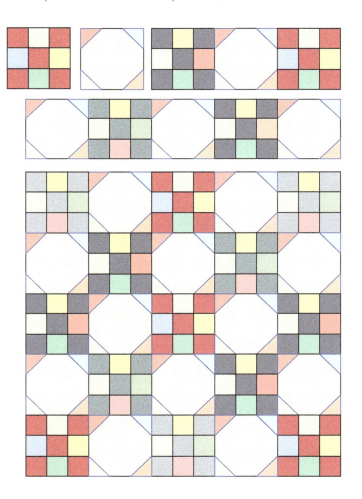

Adding the Inner Border

8 Determine the vertical measurement from top to bottom through the centre of your quilt top. It should be 42½in. Cut two side borders to this measurement (or to the measurement of your quilt if this is different to ours). Sew to the quilt and press.

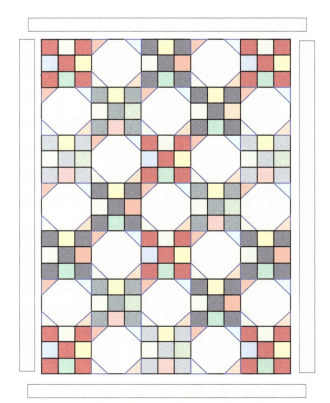

9 Determine the horizontal measurement from side to side across the centre of the quilt top. It should be 34½in – cut the top and bottom borders to this measurement (or to the measurement of your quilt if this differs). Sew to the quilt and press.

Making the Stripe Border

10 Take the four jelly roll strips and the background strip allocated for the stripe border and sew together as shown. Press seams in one direction. Cut into sixteen 2½in wide segments.

11 Sew four of the segments together into one continuous length, together with one of the extra strips from the nine-patch blocks. There will be twenty-three squares in total. This makes one side border. Repeat to make the other side border.

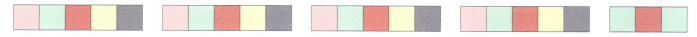

12 Sew another set of four segments together and unpick one square from one end. You need nineteen squares in total. This makes the top border. Repeat to make the bottom border.

Adding the Stripe Border

13 Pin the side borders to your quilt top, easing where necessary. Sew to the quilt. Press the work. Pin the top and bottom borders, easing where necessary. Sew to the quilt.

14 Your quilt top is now complete. Quilt as desired (see page 118 for advice) and bind to finish (see page 118).

Fairy Steps Quilt

VITAL STATISTICS
Quilt size: 47in x 54in
Setting: 5 rows of 20 blocks and four 2in wide sashing strips plus 2in border strips

REQUIREMENTS
- Half a jelly roll OR twenty 2½in strips cut across the width of the fabric
- 1.5yd (1.4m) of fabric for sashing, borders and binding
- 30in (75cm) of fabric for edging triangles

Sorting the Strips
- There is no need to do any sorting of strips at this stage.

Cutting Instructions
Jelly roll strips:
- Cut each jelly roll strip into five rectangles 2½in x 8in. You need 100.

Sashing, border and binding fabric:
- Cut thirteen 2½in wide strips *lengthways* for the sashing and borders. This will mean you don't have to have any joins in the sashing and borders.
- Set five strips aside for the binding.

Edging triangles:
- Cut six 4½in wide strips across the width of the fabric and sub-cut each into nine 4½in squares.
- Cut across both diagonals of each square to make 200 edging triangles – you will have sixteen spare.

Sewing the Quilt

1. With right sides together, sew two edging triangles to either side of a 2½in x 8in rectangle. You will see that the edging triangles are slightly larger so when sewing them to the rectangles make sure the bottom of the triangle is aligned with the rectangle as shown below. Press towards the rectangle. Repeat to make 100 of these units. For speed, chain piece all the right-hand triangles first and then rotate the unit to chain piece the left-hand triangles.

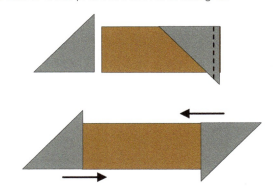

2. Take two units and with right sides together sew them together. Note how they overlap at each end which creates a straight edge when pressed open. Press open and trim the excess triangle at the back to reduce bulk.

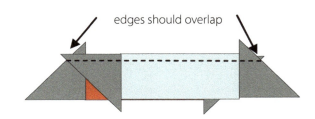

3. Sew twenty units together in this way to create one row. Repeat to make five rows of twenty units each.

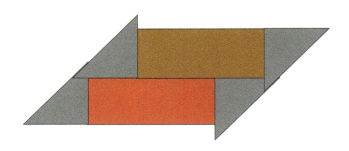

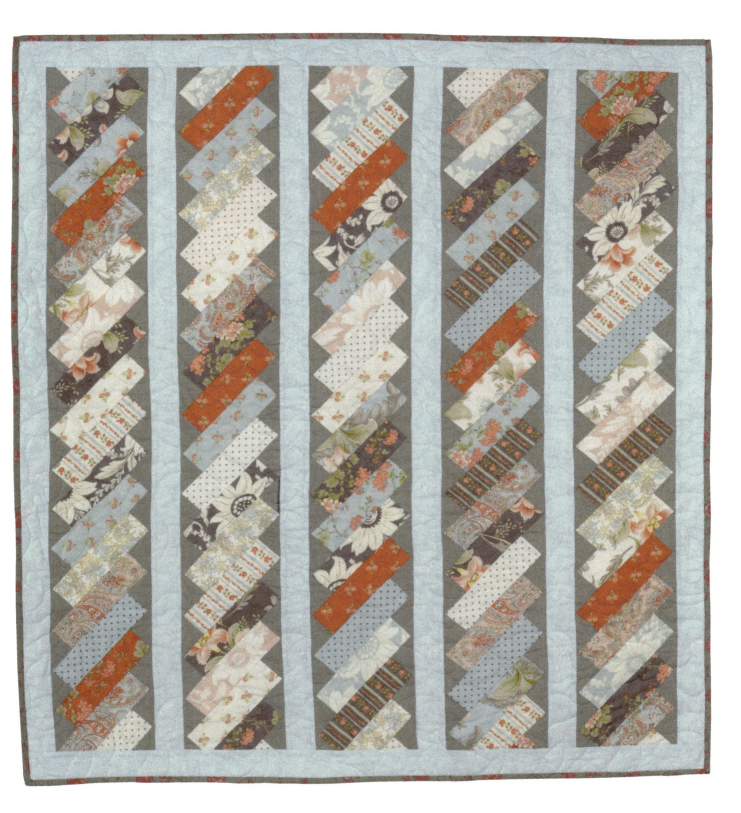

For Fairy Steps we chose a subtle aqua sashing fabric that did not dominate too much but you could create a very different effect by making the sashing fabric quite bold. Our instructions are for a binding made from border and sashing fabrics so there is no wastage, but in this instance we used 'quilter's licence' and used a darker fabric to frame our quilt. The quilt was pieced by the authors and longarm quilted by The Quilt Room.

Fairy Steps Quilt

Sewing the Sashing Strips

4 Measure the rows from the short side on the top left of the row to the short side on the bottom right of the row. Trim six sashing strips to this length. It is important that the sashing strips all measure the same. It is better to allow an inch or so extra, which can be trimmed later, than cut these sashing strips too short.

5 Pin and sew a sashing strip to the right-hand side of the five rows, easing if necessary. Press to the sashing fabric.

6 Lay out the rows and when you are happy with the arrangement and have decided which is row 1, pin and sew a sashing strip to the left-hand side of row 1, easing if necessary – see diagram below. Press towards the sashing fabric.

7 Sew the rows together pinning and easing where necessary. Press the work when all strips are added.

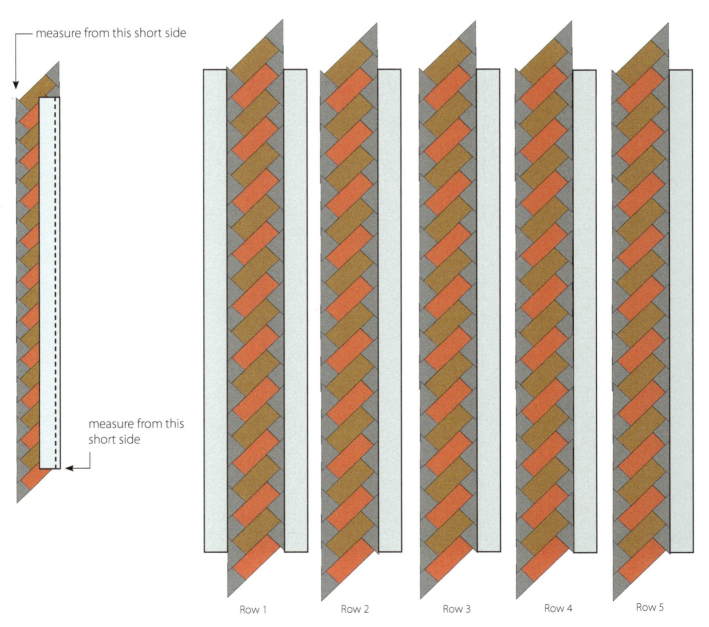

Adding the Borders and Finishing

8 Determine the horizontal measurement through the quilt centre and trim the remaining two border strips to this measurement. Pin and sew to the top and bottom of the quilt ensuring you are pinning these borders at right angles to the side borders. Trim excess fabric and press.

9 Your quilt top is now complete. Quilt as desired (see page 118 for advice) and bind to finish (page 118).

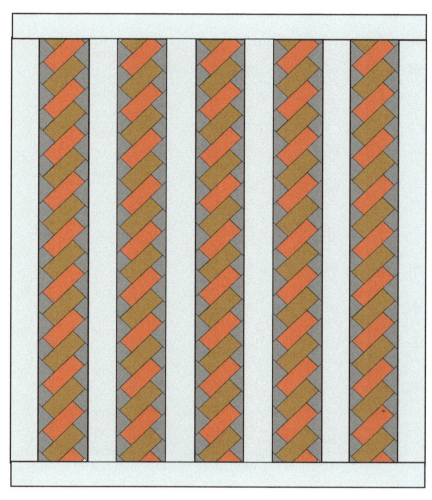

Fairy Steps Quilt 67

Chapter Seven
Spinning Tops and Train Tracks

Any quilt made from these colourful fabrics would brighten any room and the two quilts in this chapter are certain to do that. The fabric range is called Spring Fever from Me & My Sister and has an unusual mix of colours – gorgeous apple greens, pinks, yellows and mauve – oh yes, and tangerine as well!

After much debate we chose our fabrics for the Spinning Tops quilt to be a mix of apple green and mauve for colour A and pink and tangerine for colour B. This isn't the most usual colour combination but see how great it looks – very vibrant and dynamic.

The design for the Train Tracks quilt uses the same colourful fabrics but creates a very different look with its combination of four-patch blocks and snowball blocks.

Two simple blocks, a four-patch and a snowball, make up the design for Train Tracks – very easy and very colourful.

Spinning Tops Quilt

VITAL STATISTICS
Quilt size: 47in x 58in
Block size: 11in square
Number of blocks: 12
Setting: 3 x 4 blocks plus 2in inner border and 5¼in stripe border

REQUIREMENTS
- Half a jelly roll OR twenty 2½in wide strips cut across the width of the fabric
- 16in (40cm) of accent fabric for colour A
- 16in (40cm) of accent fabric for colour B
- 1¾yd (1.5m) of neutral and border fabric
- 20in (50cm) of fabric for binding

Sorting the Strips
- Divide the jelly roll strips into ten strips of colour A and ten strips of colour B. You don't have to have all the same colour in one pile but there should be a distinction between the two piles.
- Take the ten strips from colour A and pair them up to form five pairs.
- Do the same with the ten colour B strips. You now have five pairs of colour A and five pairs of colour B.

Cutting Instructions
Fabric Accent A:
- Cut five 2½in wide strips across the width of the fabric.

Fabric Accent B:
- Cut five 2½in wide strips across the width of the fabric.

Neutral and border fabric:
- Cut fourteen 2½in wide strips across the width of the fabric. Set four aside for the inner border.
- Cut two 5¾in wide strips across the width of the fabric. Sub-cut one 5¾in wide strip into four rectangles 5¾in x 10½in.
Sub-cut one 5¾in wide strip into four squares 5¾in x 5¾in. These squares and rectangles are for outer border corners and will be trimmed to the exact size later.

Binding:
- Cut six 2½in wide strips across the width of the fabric.

Making the Blocks
1. Take one of the five pairs from the colour A jelly roll strips and, with right sides together, sew down the long side. Press the work.

2. Sew a neutral strip to the top of the strip unit and sew a colour A accent strip to the bottom of the strip unit. Press the work.

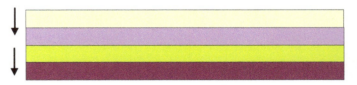

3. Take a quilting ruler, which has a marked 60 degree line, and place the 60 degree line along the bottom of the strip unit and position the bottom left-hand corner of the ruler in the bottom left-hand corner of the strip unit. Double check that your ruler is in the correct position before cutting. Cut along the left-hand edge of the ruler.

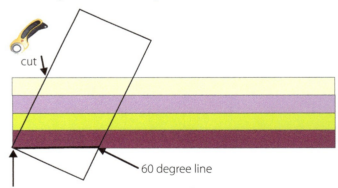

Position corner of ruler in bottom left-hand corner

4. Cut five 6in segments from the strip unit, checking that the 60 degree mark on your ruler is always aligned with the bottom of the strip unit. Re-trim the cut end if necessary as you want to be sure you are always cutting at a 60 degree angle. The offcuts from each end are spare.

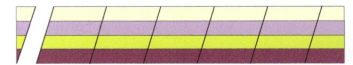

This vibrant design is sure to attract attention! There were a few offcuts spare and if we'd had time we would have used them in a pieced backing. This is a good option if you ever do have some pieces spare, as an interesting backing always enhances a quilt. The quilt was pieced by the authors and longarm quilted by The Quilt Room.

5 Take one segment and rotate it 90 degrees so the neutral strip is on the right. Line up the quilting ruler so that the top right-hand corner is on the top right-hand corner of the strip unit. Cut along the right-hand edge of the ruler.

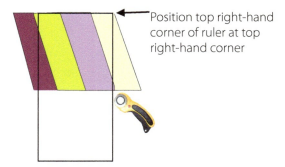

Position top right-hand corner of ruler at top right-hand corner

6 Rotate the segment 180 degrees so that the neutral strip is on the left side. Using the horizontal and vertical lines of your quilting ruler, cut to create a 6in square. Set the two offcuts aside, keeping them as a pair to be used later in the outer border. They may be slightly different in size but don't worry as they will be trimmed later. Repeat with the other four segments of the strip unit.

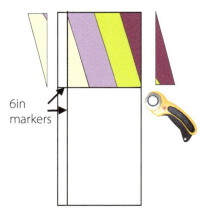

6in markers

7 Take four of the segments and sew them together to form a colour A block as shown below. The fifth segment will be used in a colour A block which is made up of mixed segments. Handle the segments carefully as the edges are on the bias and when pressing do not use steam.

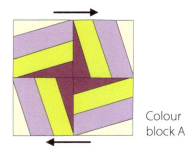

Colour block A

8 Repeat steps 1–7 with the other pairs of strips from colour A. You need six colour A blocks and the sixth block is made up of four of the spare segments. You will have one segment spare.

9 Repeat steps 1–8 with all the pairs of strips from the colour B.

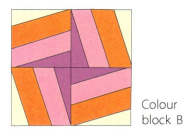

Colour block B

Assembling the Quilt

10 Lay out the blocks into rows alternating the colour A and colour B blocks. Make four rows of three blocks. When you are happy with the layout, sew the blocks into rows and then sew the rows together. Pin at every seam intersection to ensure matching seams. Press carefully making sure you do not stretch the outside edges and do not use steam.

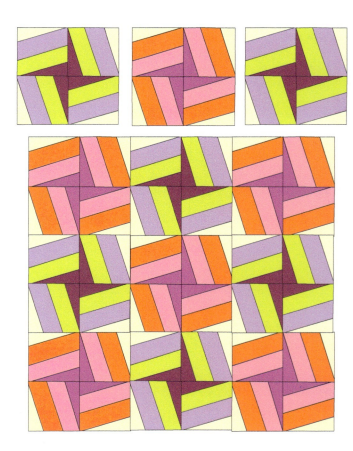

Adding the Inner Border

11 Join the four 2½in border strips into a continuous length and, referring to the instructions on page 117, sew the inner borders to the quilt. Press to the border fabric.

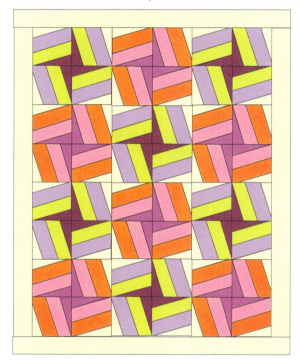

Adding the Outer Border

12 Take a pair of the triangle offcuts and sew into a rectangle. Note that when sewing the triangles together they will overlap at each end. Press to the darker fabric. Don't worry if any of your triangles are slightly different in size and don't bother to trim the dog ears as the borders will be trimmed after the rectangles are sewn together.

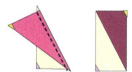

13 Repeat with the rest of the triangle offcuts to form twenty-four rectangles from colour A and twenty-four rectangles from colour B. You will have two spare.

14 To make the side borders, sew fourteen rectangles together alternating colour A and colour B rectangles and sew a 5¾in x 5¾in neutral square to each end. Repeat to form the other side border. Press seams in one direction.

15 To make the top and bottom borders, sew ten rectangles together and then sew a 5¾in x 10½in rectangle to each end. Press seams in one direction.

16 At this stage, using your quilting ruler and rotary cutter, trim the four borders down the long edges to ensure they are straight. This will also trim off all the dog ears. Don't trim the short ends to size yet.

Adding the Borders and Finishing

17 Determine the vertical measurement from top to bottom through the centre of your quilt top. Trim the two side borders to this measurement, trimming some from each end so the border pattern is central. Pin and sew to the quilt. Press the work.

18 Determine the horizontal measurement from side to side across the centre of the quilt top. Trim the top and bottom borders to this measurement, trimming some from each end so the border pattern is central. Pin and sew to the quilt. Press the work.

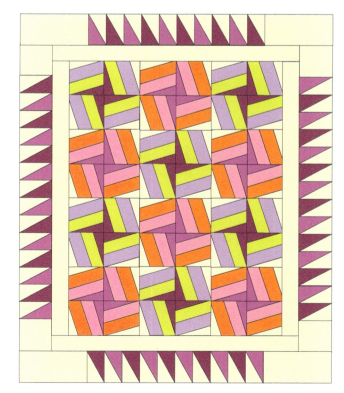

19 Your quilt top is now complete. Quilt as desired (see page 118 for advice) and bind to finish (page 118).

Spinning Tops Quilt

Train Tracks Quilt

VITAL STATISTICS
Quilt size: 36in x 52in
Block size: 8in square
Number of blocks: 24
Setting: 4 x 6 blocks plus 2in border

REQUIREMENTS
- Half a jelly roll OR twenty 2½in strips cut across the width of the fabric
- 32in (80cm) of fabric for background
- 16in (40cm) of fabric for border
- 16in (40cm) of fabric for binding

Sorting the Strips
- Choose six jelly roll strips for the corners of the snowball units. This could be in one colourway as we have done or an assortment.
- Choose twelve jelly roll strips to be used in the four-patch units. Two strips will be spare.

Cutting Instructions
Jelly roll strips:
- Take the six jelly roll strips allocated for the corners of the snowball units and sub-cut each strip into sixteen 2½in squares to make ninety-six in total. You can layer up the strips before cutting but make sure you don't cut through too many as you will lose accuracy. It is better to have two layers of three strips and butt them up against each other and then cut your squares.

Background fabric:
- Cut six 4½in strips across the width of the fabric. Sub-cut each strip into eight 4½in squares to make forty-eight in total.

Border fabric:
- Cut five 2½in wide strips across the width of the fabric.

Binding fabric:
- Cut five 2½in wide strips across the width of the fabric.

Making the Four-Patch Units

1. Cut the twelve assorted strips allocated for the four-patch units in half to create twenty-four rectangles 2½in x 21in. This ensures more variety in the four-patch units.

2. Take two contrasting 2½in x 21in rectangles and lay right sides together. Sew down the long side. Open and press to the darker fabric. Repeat with the remaining assorted 2½in x 21in rectangles to make a total of twelve strip units, chain piecing for speed. Open and press to the darker fabric.

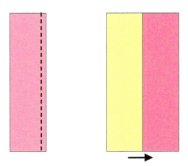

3. With right sides together, lay one strip unit on top of another with the light strip on the top of one unit and on the bottom of another, ensuring that the centre seams are in alignment. Sub-cut into eight 2½in wide segments.

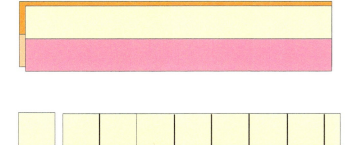

4. Carefully keeping the pairs together, sew down the long side as shown, pinning at the seam intersection to ensure a perfect match. The seams will nest together nicely as they are pressed in different directions. Chain piece for speed. Press open to form eight four-patch units.

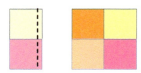

5. Repeat with the remaining strips to make a total of forty-eight four-patch units.

Making the Snowball Units

6. Draw a diagonal line from corner to corner on the wrong side of the 2½in squares allocated for the snowball unit corners.

7. With right sides together, lay a marked square on one corner of a 4½in background square, aligning the outer edges. Sew across the diagonal, using the marked diagonal line as the stitching line. After a while you may find you do not need to draw the line as it is not difficult to judge the sewing line. Alternatively, you can mark the line with a fold.

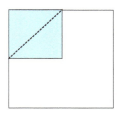

8. Flip the square over and press towards the outside of the block. Trim the excess fabric from the snowball corner but do not trim the background fabric. Although this creates a little more bulk, the background fabric helps keep the patchwork in shape.

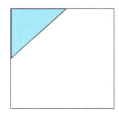

9. Repeat on the opposite corner. Repeat the process to make forty-eight snowball units.

Making the Block

10 Sew a four-patch unit to a snowball unit as shown right. Press towards the snowball unit. Repeat with another four-patch unit and a snowball unit. Rotate one unit 180 degrees and sew the two units together to form one block. Pin at every seam intersection to ensure a perfect match.

11 Repeat with all forty-eight four-patch units and forty-eight snowball units to create twenty-four blocks.

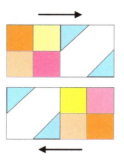

Assembling the Blocks

12 Lay out the blocks as shown in the diagram, right. Note that the lower half of the quilt is the same as the top half but rotated 180 degrees. When you are happy with the arrangement sew the blocks into rows and then sew the rows together, pinning at every seam intersection to ensure a perfect match. Press the seams in alternate rows in opposite directions and you will find the seams nest together nicely when joining the rows.

Train Tracks Quilt

Adding the Borders and Finishing

13 Join the five 2½in border strips into one continuous length and, referring to the diagram here and the instructions on page 117, add the borders to the quilt.

14 Your quilt top is now complete. Quilt as desired (see page 118 for advice) and bind to finish (see page 118).

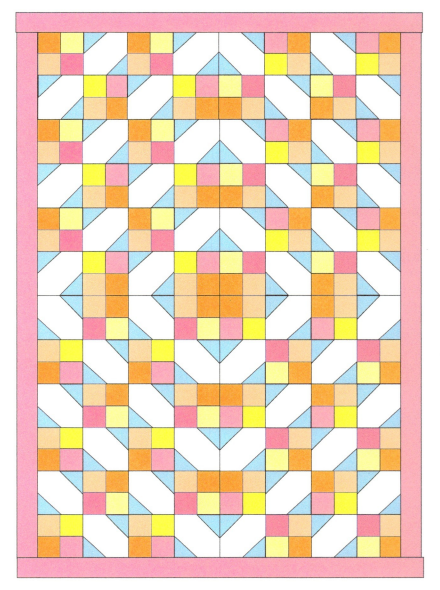

Our original idea for this quilt was to have the background and border from the same fabric but in the end we liked the added colour of the pink border – though we can't imagine why we needed to add more colour! The quilt was pieced by the authors and longarm quilted by The Quilt Room.

Chapter Eight
Kaleidoscope and Jigsaw

We originally used a Fig Tree jelly roll for these quilts and they looked great – job done! Then this range called Rural Jardin from French General arrived and we couldn't resist trying the fabrics. It's every quilter's dilemma – so many lovely fabrics and you just don't know which to use! You can, however, see our quilts using the Fig Tree range on page 126 – so this time it's a win-win situation all round.

For Kaleidoscope we pulled out the dark reds and dark blues from the range as we wanted this to be a dynamic looking quilt. We didn't have many true neutrals in the range so we chose the pale blue as our neutral. Be guided by what is in your jelly roll – the fabrics all blend together beautifully which should give you confidence. The Jigsaw quilt looks quite different, thanks to the gentler, calmer colour scheme.

Once the dark reds and dark blues came out of the jelly roll this left the lovely softer colours for the Jigsaw quilt. It is made from the traditional sawtooth block, which has always been one of our favourites.

Kaleidoscope Quilt

VITAL STATISTICS
Quilt size: 34in x 44in
Block size: 10in square
Number of blocks: 12
Setting: 3 x 4 blocks plus 2in border

REQUIREMENTS
- Half a jelly roll OR twenty 2½in strips cut across the width of the fabric
- 12in (30cm) of fabric to match colour A (dark red)
- 12in (30cm) of fabric for centre squares
- 12in (30cm) of fabric for border
- 12in (30cm) of fabric for binding
- Omnigrid 96 ruler or other speciality ruler

Sorting the Strips
- Sort your jelly roll strips as follows (one of the strips will be spare):
  five colour A (dark red);
  four colour B (blue);
  two colour C (aqua);
  two colour D (light red);
  six neutral strips.

Cutting Instructions
Extra fabric – colour A (dark red):
- Cut six strips 1½in wide across the width of the fabric.
- Sub-cut each strip into eight rectangles 1½in x 4½in. You need forty-eight in total.

Extra fabric – centre squares:
- Cut two strips 4½in wide across the width of the fabric.
- Sub-cut each strip into six squares 4½in x 4½in. You need twelve in total.

Jelly roll strips:
- Take three of the colour A (dark red) strips and cut each into sixteen squares 2½in x 2½in. You need forty-eight in total. Leave the other two dark red strips uncut.
- Take two colour B (blue) strips and trim them to measure 1½in x 42in. Sub-cut each strip into twenty-four squares 1½in x 1½in. You need forty-eight in total. Leave the other two blue strips uncut.
- Take the two colour D (light red) strips and cut each into twelve squares 2½in x 2½in. You need twenty-four in total.
- Take four of the neutral strips and cut each into six rectangles 2½in x 6½in. You need twenty-four in total. Leave the other two neutral strips uncut.

Border:
- Cut four 2½in wide strips across the width of the fabric.

Binding:
- Cut four 2½in wide strips across the width of the fabric.

Making the Half-Square Triangle Units

1. Take two colour C (aqua) jelly roll strips and two colour B (blue) jelly roll strips. Press one aqua and one blue strip right sides together ensuring that they are exactly one on top of the other. The pressing will help to hold the two strips together.

2. Lay out on a cutting mat and trim the selvedge on the left side. Position the Omnigrid 96 as shown in the diagram below, lining up the 2in mark at the bottom edge of the strips and cut the first triangle. You will notice that the cut out triangle has a flat top. This would just have been a dog ear you needed to cut off – so it is saving you time!

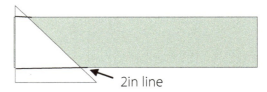

3. Rotate the ruler 180 degrees to the right as shown and cut the next triangle. Continue along the strip. You need twenty-four sets of triangles from one strip.

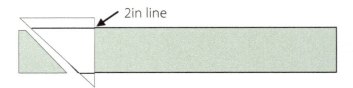

This is a great quilt – so dynamic and interesting with the focus on those lovely dark reds and dark blues. It's easy to piece too. The quilt was pieced by the authors and longarm quilted by The Quilt Room.

Kaleidoscope Quilt

4. Sew along the diagonals to form twenty-four half-square triangles, chain piecing for speed (see page 113). Trim all dog ears and then press open with the seams pressed towards the aqua (lighter) fabric. Repeat with the other aqua and blue strips to make forty-eight aqua and blue half-square triangles in total.

5. Take two colour A (dark red) and two neutral jelly roll strips and repeat steps 1–4 to make forty-eight dark red and neutral half-square triangle units. Press towards the darker fabric.

Making the Centres of Blocks A and B

6. Take two dark red 1½in x 4½in rectangles and sew to either side of a 4½in centre square. Press as shown. Repeat to create twelve of these units.

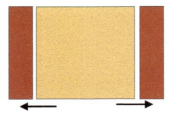

7. Take two colour B (blue) 1½in squares and sew to either side of a colour A (dark red) 1½in x 4½in rectangle. Press as shown. Repeat to create twenty-four of these units.

8. Sew these units to the top and bottom, pinning at every seam intersection to ensure a perfect match. Press as shown. Repeat to make twelve centre units.

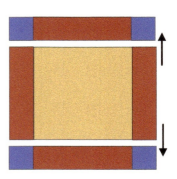

Making Block A

9. Sew two dark red and neutral half-square triangle units to either side of a light red square as shown. Press the work. Repeat to make four of these units.

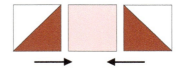

10. Sew two of these units to either side of a centre square as shown and then press the work.

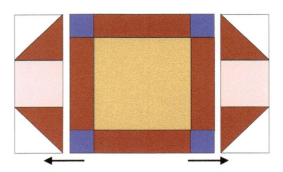

11. Sew two aqua and blue half-square triangle units to either side of the other two units. Press the work.

12. Sew these to the top and bottom to create Block A, pinning at every seam intersection to ensure a perfect match. Repeat to make six of Block A.

Block A

Making Block B

13. Take one dark red 2½in square and lay it right sides together on a 2½in x 6½in neutral rectangle as shown. Sew across the diagonal. Folding your square to mark the diagonal line will help keep your stitching accurate.

14. Flip the square over and press towards the dark red fabric. Trim the excess dark red fabric but do not trim the neutral fabric. Although this creates a little more bulk, this neutral rectangle keeps your units in shape.

15. Place a second dark red 2½in square and lay it on the other side as shown and sew across the diagonal. Flip the square over and press the work. Trim the excess dark red fabric. Repeat to make four units.

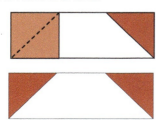

16. Sew two of these units to either side of a centre square and press as shown in the diagram.

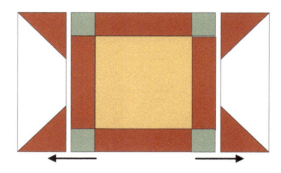

17. Sew two aqua and blue half-square triangle units to either side of the other two units and press as shown.

18. Now sew these units to the top and bottom, as shown below, to create Block B, pinning at every seam intersection to ensure a perfect match. Press the work. Repeat this process to make six of Block B.

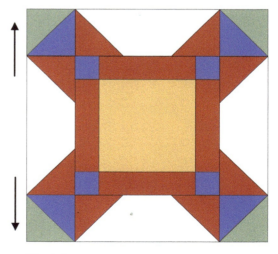

Block B

Kaleidoscope Quilt 85

Assembling the Quilt

19 Referring to the diagram below lay out the blocks into rows, alternating Block A and Block B. Make four rows of three blocks. When you are happy with the layout, sew the blocks into rows, pressing the rows in alternate directions. Sew the rows together, pinning at every seam intersection to ensure you have matching seams. Press the work.

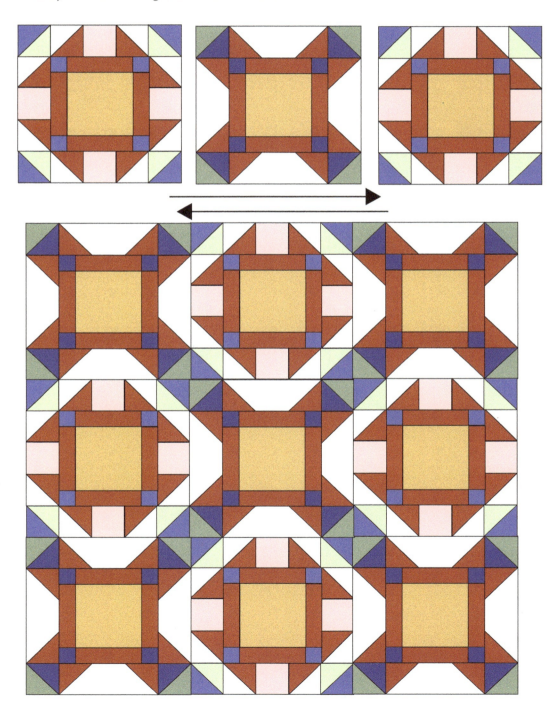

Adding the Borders and Finishing

20 Referring to the diagram here and the instructions on page 117, add the borders to the quilt. Add the side borders first and you won't need to join the border strips into a continuous length.

21 Your quilt top is now complete. Quilt as desired (see page 118 for advice) and bind to finish (page 118).

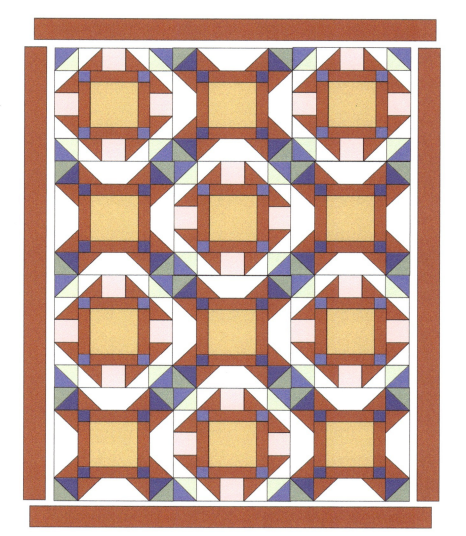

Jigsaw Quilt

VITAL STATISTICS
Quilt size: 36in x 42in
Block size: 6in square
Number of blocks: 30
Setting: 5 x 6 plus 2in sawtooth top and left border and 2in outer border

REQUIREMENTS
- Half a jelly roll OR twenty 2½in strips cut across the width of the fabric
- 1yd (1m) of fabric for background
- 16in (40cm) of fabric for binding
- Omnigrid 96 ruler or other speciality ruler

Sorting the Strips
- Choose sixteen strips and make up eight pairs of strips in similar colours.
- Choose one strip to be added to the outer border. The other three are spare.

Cutting Instructions
- Take the sixteen jelly roll strips which are now in pairs and cut into the following rectangles, keeping the 12in and 21in rectangles from the pairs together:
2½in x 12in rectangle;
2½in x 21in rectangle;
2½in x 9in rectangle and set these aside for the outer border.
- Take the other strip and cut into four rectangles 2½in x 9in and set aside with the others already cut for the outer border.

Background fabric:
- Cut three 4½in wide strips across the width of the fabric.
- Sub-cut each strip into three rectangles 4½in x 12in. You need eight in total – one will be spare.
- Cut eight 2½in wide strips across the width of the fabric and cut each in half to create two rectangles 2½in x 21in. You need sixteen in total.

Binding fabric:
- Cut four 2½in strips across the width of the fabric.

Making the Blocks

1. Working with one pair of jelly roll strips, take a pair of 2½in x 12in rectangles and, with right sides together, sew down the long side. Open and press the work.

2. With right sides together, lay this unit on a 4½in x 12in background strip aligning the edges and press to hold in place. It is important to always have the background strip on the bottom. Trim the left-hand selvedge. Position the Omnigrid 96, lining up the 4in mark at the bottom edge of the strips and cut the first triangle. You will notice that the cut-out triangle has a flat top. This would just have been a dog ear you needed to cut off – so it is saving you time!

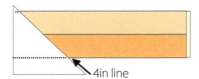

3. Rotate the ruler 180 degrees to the right as shown and cut the next triangle. Repeat to cut four pairs of triangles.

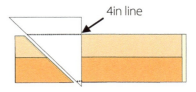

4. Sew along the diagonals to form four 4in half-square triangle units. Trim all dog ears and press open. You will have two each of two different 4in half-square triangle units.

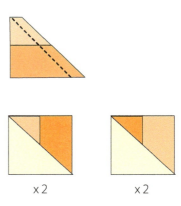

5 Still working with the same pair of jelly roll strips, take one of the 2½in x 21in jelly roll rectangles and a 2½in x 21in background rectangle and lay them right sides together – it doesn't matter which fabric is on top. Press to hold in place. Using the Omnigrid 96 as before but this time using the 2in line for cutting, cut twelve pairs of triangles. Sew along the diagonals to form twelve 2in half-square triangle units and press open.

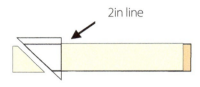

6 Repeat with the other 2½in x 21in jelly roll strip to make a further twelve 2in half-square triangle units. Press open. You now have twenty-four in total.

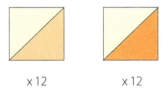

x 12   x 12

7 Take five identical 2in half-square triangle units. Sew two together as shown and sew to the bottom of a 4in half-square triangle unit. Press as shown. We chose to have our fabrics match the base of the large triangle but you could have them match the top or mix them all up – the choice is yours.

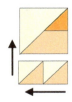

8 Sew the other three 2in half-square triangle units together as shown. Press and sew to the right-hand side, pinning at the seam intersections to ensure a perfect match.

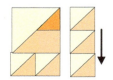

9 Repeat steps 1–8 to make four blocks as shown. Set aside the extra four 2in half-square triangle units for the sawtooth border.

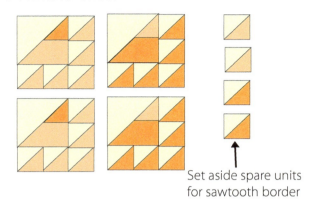

Set aside spare units for sawtooth border

10 Repeat steps 1–9 to use all your eight pairs of jelly roll strips. You only need thirty blocks so when working with the final pair of strips, piece only two blocks. Leave the remaining two blocks unpieced, as two of the 2in half-square triangle units are needed for the sawtooth border. You need a total of thirty-four 2in half-square triangle units for the sawtooth border.

Assembling the Quilt

11 Lay out the blocks into six rows of five blocks. When you are happy with the layout, sew the blocks into rows and then sew the rows together. Pin at every seam intersection to ensure matching seams. Be prepared when sewing the blocks together to re-press the occasional seam to ensure they nest together nicely and create less bulk.

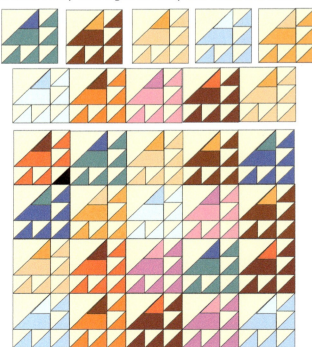

Adding the Sawtooth Border

12 The sawtooth border is only needed on the top and left-hand side of the quilt. Sew eighteen 2in half-square triangles together as shown. These are for the left-hand sawtooth border.

13 Sew sixteen 2in half-square triangles together as shown. These are for the top sawtooth border.

14 Sew the left-hand side border on first, pinning and easing if necessary and then press the work carefully. Sew on the top border, pinning and easing if necessary. Press the work.

Note: when sewing pieced units together that have points, you don't want to lose those points, so make sure you sew to the *right* of the points and you won't lose the nice sharp tips.

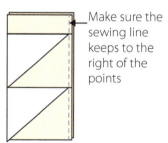

Make sure the sewing line keeps to the right of the points

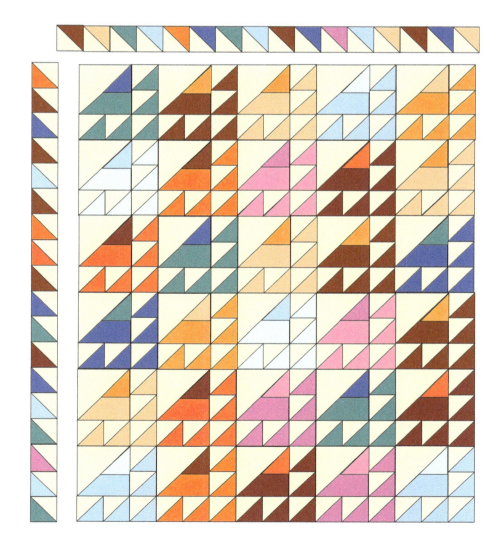

Jigsaw Quilt

Adding the Outer Border

15 Join five 2½in x 9in rectangles into a continuous length and press the work. Repeat to make four of these. Referring to the instructions on page 117, add the outer border to the quilt, adding the side borders first and trimming where required.

16 Your quilt top is now complete. Quilt as desired (see page 118 for advice) and bind to finish (see page 118).

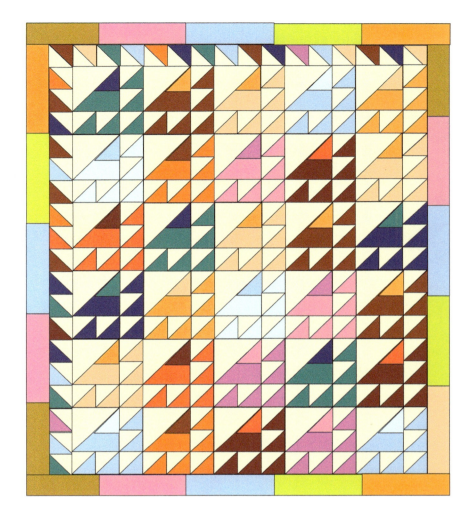

92 Jigsaw Quilt

What a gorgeous quilt this is. It took us a while to fathom out how to create this design using jelly roll strips but it turned out to be very simple – it always is once you know how! The quilt was pieced by the authors and longarm quilted by The Quilt Room.

Jigsaw Quilt

Chapter Nine
Teddy Bears and Loving Hearts

Despite being from the same jelly roll these two quilts look very different. We split the jelly roll, using the bolder shades for the Teddy Bear quilt and the more delicate pastels for the Loving Hearts quilt. We didn't want our teddies to be too dark but wanted some brown in there somewhere and this super range from Moda's Three Sisters works well. The quilt has lots of flip-over corners but these are not difficult and the charming result is well worth the extra time.

The Loving Hearts quilt is perfect for a loved one, and just about any colours would work well. We set the blocks on point with narrow sashing strips to make the hearts stand out.

The Loving Hearts quilt is a wonderful way to show someone how much you care. With its subtle colouring it's hard to believe it was created from the same jelly roll as the Teddy Bears quilt.

Teddy Bears Quilt

VITAL STATISTICS
Quilt size: 46in x 46in
Block size: 20in square
Number of blocks: 4
Setting: 2 x 2 plus 2in sashing

REQUIREMENTS
- Half a jelly roll OR twenty 2½in strips cut across the width of the fabric
- 1¼yd (1m) of fabric for background
- One fat quarter of coordinating red fabric
- 20in (50cm) of fabric for sashing
- Spare jelly roll strips for binding

Sorting the Strips
- Choose six reds for the clothes.
- Choose two light browns for the bears.
- Choose four dark browns for the bears.
- Choose one strip for the bow ties.
- Choose two strips for the sashing squares.
- Choose five strips for the binding.

Cutting Instructions
Red clothes:
- From the six red strips for the clothing cut as follows: two strips – cut each into eight rectangles 2½in x 4½in; two strips – cut each into two rectangles 2½in x 16½in, plus one 2½in x 6½in rectangle; one strip – cut into six rectangles 2½in x 6½in; one strip – cut into eight 2½in x 2½in squares.

Light brown for bears:
- Cut the two strips in half to make four rectangles 2½in x 21in.
- Sub-cut the half strips into one 2½in x 8½in rectangle and two 2½in x 4½in rectangles, keeping the pieces from each half strip together.

Dark brown for bears:
- Cut each strip into seven 2½in x 4½in rectangles and four 2½in x 2½in squares.
- Keep the fabrics from each strip in four separate piles.

Bow ties:
- Cut the strip into sixteen 2½in x 2½in squares.

Sashing squares:
- Cut the two strips into twenty-five 2½in squares.

Background fabric:
- Cut twelve 2½in wide strips across the fabric width.
- Sub-cut four strips into sixteen 2½in x 2½in squares each, to make a total of sixty-four.
- Sub-cut two strips into eight 2½in x 4½in rectangles each, to make a total of sixteen.
- Sub-cut six strips into two 2½in x 16½in rectangles each, to make a total of twelve.
- Cut one 4½in wide strip across the width of the fabric and sub-cut into eight 4½in x 4½in squares.

Co-ordinating fat quarter:
- Cut eight 4½in x 4½in squares.

Sashing:
- Cut six 2½in wide strips across the width of the fabric.
- Sub-cut each strip in half to make twelve rectangles 2½in x 21in. These will need to be trimmed to size later.

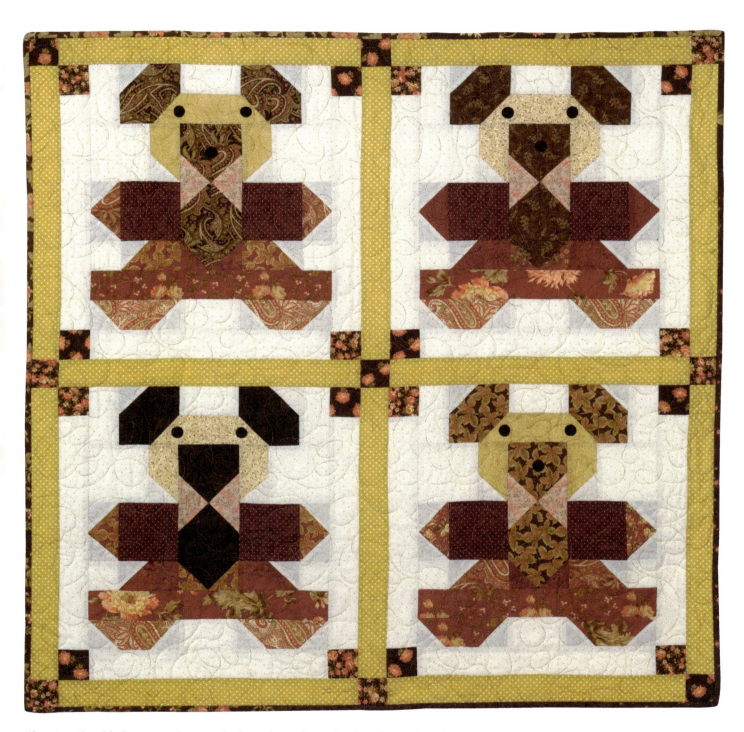

The pieced teddy bears on this cute little quilt are bound to be a hit with children and grown-ups alike. The design uses bold reds and browns from the jelly roll but the teddies would look good in any colour. You could also make more blocks for a bigger quilt. The quilt was pieced by the authors and longarm quilted by The Quilt Room.

Teddy Bears Quilt

Making a Teddy Bear Block

1 Work with one pile of light brown strips and one pile of dark brown strips plus an assortment of reds for each block.

2 Unit A: Take one 2½in background square and lay it right sides together on a 2½in x 4½in dark brown rectangle. Sew across the diagonal, as shown in the diagram below. If it helps, draw the diagonal line in first or make a fold to mark the stitching line. Flip the square over and press towards the background fabric and then trim the excess background fabric.

3 Take another 2½in background square and sew it to a second dark brown rectangle on the opposite side, as shown.

4 Sew these two units to either side of a 2½in x 4½in background rectangle. Press in the directions shown.

5 Take two dark brown 2½in squares and using the same flip and sew method as before sew them to both ends of a light brown 2½in x 8½in rectangle. Press and trim the dark brown excess fabric.

6 Sew a 2½in dark brown square to both ends. Press away from the dark brown square.

7 Sew these two units together, pinning at every seam intersection to ensure a perfect match. Sew a 2½in x 4½in background rectangle to both sides to form unit A. Press to the background rectangles.

Unit A

8 Unit B: Using the same flip and sew method as before, sew a 2½in background square to two light brown 2½in x 4½in rectangles. Press and trim excess background fabric.

9 Using the same flip and sew method, sew two 2½in bow tie squares to the bottom of two dark brown 2½in x 4½in rectangles as shown. Press and trim the excess bow tie fabric. Sew the rectangles together and press.

10 Sew these units together with a 4½in background square at each end to form unit B. Press as shown.

Unit B

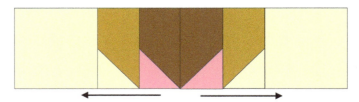

11. **Unit C:** Using the same flip and sew method, sew two 2½in bow tie squares to the top of two dark brown 2½in x 4½in rectangles as shown. Press and trim the excess bow tie fabric. Sew the rectangles together.

12. Using the same flip and sew method, sew two 2½in background squares to both ends of two red 2½in x 4½in rectangles as shown. Press and trim the excess background fabric.

13. Assemble unit C by sewing together as shown with the 4½in red squares. Press as shown.

Unit C

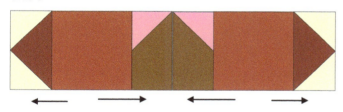

14. **Unit D:** Using the same flip and sew method, sew a 2½in background square to one end of two red 2½in x 4½in rectangles as shown. Press and trim the excess background fabric.

15. Using the same flip and sew method, sew two 2½in red squares to both ends of a dark brown 2½in x 4½in rectangle as shown. Press and trim the excess red fabric.

16. Assemble unit D by sewing together as in the diagram below and sewing a 2½in background square to both ends. Press as shown.

Unit D

17. **Unit E:** Using the same flip and sew method, sew two 2½in background squares to both ends of two red 2½in x 6½in rectangles as shown. Press and trim the excess background fabric.

18. Assemble unit E as shown in the diagram, sewing the units to both sides of a 2½in x 4½in background rectangle and then sewing a red 2½in x 16½in rectangle to the top.

Unit E

Assembling the Block

19 Sew units A, B, C, D and E together to make the teddy bear block as shown below, pinning at every seam intersection to ensure a perfect match. Press the work.

20 Sew a 2½in sashing square to one end of a 2½in x 16½in background rectangle. Repeat and sew one to either side of the block. Sew a sashing square to both sides of a 2½in x 16½in background rectangle and sew to the bottom of the block. Repeat to make four teddy bear blocks.

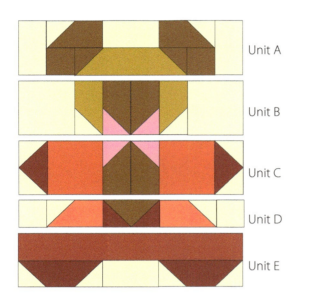

Unit A
Unit B
Unit C
Unit D
Unit E

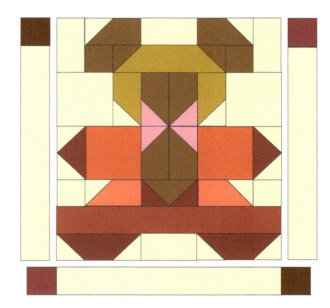

Assembling the Quilt

21 Measure the teddy bear block – it should be 20½in square. Trim your sashing strips to this size or the size of your block if it differs. Sew two sashing strips with a sashing square in the centre and a sashing square at both ends. Make three of these.

22 Sew two teddy bear blocks with a sashing strip in the centre and a sashing strip on either side, pinning and easing if necessary. Repeat this sashing with the other two teddy bear blocks.

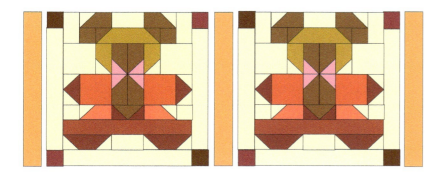

23 Now sew your quilt together as shown in the diagram right, pinning and easing if necessary.

24 Embroider eyes and noses on your bears, or do as we did and appliqué small circles approximately 1¾in in diameter – see page 115 for fusible web appliqué instructions. Buttons would not be recommended for a child's quilt.

25 Your quilt is now complete. Quilt as desired (see page 118 for advice) and bind to finish (see page 118). For the binding cut each of the jelly roll strips allocated for the binding into four rectangles 2½in x 10½in. Sew them together in a continuous length, alternating the fabrics as this will give a scrappier effect for the binding.

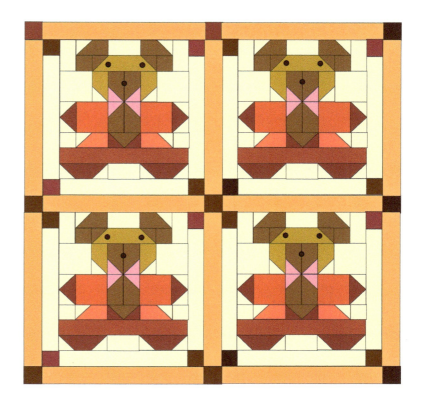

Loving Hearts Quilt

VITAL STATISTICS
Quilt size: 49in x 49in
Block size: 10in square
Number of blocks: 13
Setting: on point, with each block having a ¾in wide frame

REQUIREMENTS
- Half a jelly roll OR twenty 2½in wide strips cut across the width of the fabric
- 16in (40cm) of fabric for the outer block frames
- 1½yd (1.3m) of fabric for background and setting triangles
- 16in (40cm) of fabric for binding

Sorting the Strips
- Choose four strips to be the frames of the inner four blocks.
- The remaining sixteen will be used for the blocks.

Cutting Instructions
Jelly roll strips:
- Take the four strips allocated for the frames of the inner blocks and from each cut one rectangle 2½in x 10½in and one rectangle 2½in x 12in.
- Sub-cut each in half lengthways to create two rectangles 1¼in x 10½in and two rectangles 1¼in x 12in. Keep the rectangles from each strip together.
- From the sixteen strips allocated for the blocks, choose thirteen and from each strip cut four rectangles 2½in x 6½in, keeping the four rectangles from the same fabric together. The balance of the strips (approximately 16in) will be used to make the nine-patch blocks.
- Leave the remaining three strips uncut to make the nine-patch blocks.

Outer block frames:
- Cut eleven 1¼in wide strips across the width of the fabric.
- Sub-cut five strips into four rectangles 1¼in x 10½in to make twenty. You need eighteen so two will be spare.
- Sub-cut six strips into three rectangles 1¼in x 12in to make eighteen.

Background and setting triangle fabric:
- Cut two 4½in wide strips across the width of the fabric. Sub-cut each into eight 4½in squares. You need thirteen (three will be spare).
- Cut four 2½in wide strips across the width of the fabric. Sub-cut each into sixteen 2½in squares. You need fifty-two (twelve will be spare).
- Cut one 18in wide strip across the width of the fabric. Sub-cut into two 18in squares. Cut across both diagonals to form eight setting triangles.

 18in square

- Cut one 10in wide strip across the width of the fabric. Sub-cut into two 10in squares. Cut across one diagonal of these squares to form four corner triangles. Cutting the setting and corner triangles in this way ensures that there are no bias edges on the outside of your quilt.

 10in square

Binding:
- Cut five 2½in wide strips across the width of the fabric.

Making the Nine-Patch Blocks

1 Take the three uncut jelly roll strips allocated for the nine-patch blocks and sew them together as shown in the diagram below. Press seams in one direction. Trim the selvedge and cut into sixteen 2½in wide segments.

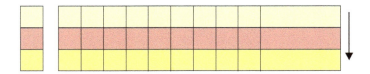

2 Take three of the 16in lengths allocated for the nine-patch blocks and sew them together. Press the seams in one direction. Repeat to make four of these units in total and cut each into six 2½in segments to make twenty-four. You will have one 16in length spare.

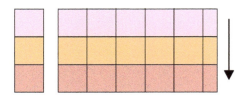

3 You have forty 2½in segments in total. Select three segments and sew together to make one nine-patch block. Repeat to make thirteen nine-patch blocks, trying to ensure you don't put the same fabrics next to each other and turning the segments to ensure the seams nest together nicely. You will have one segment spare.

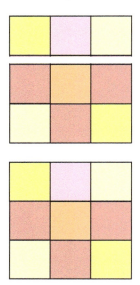

Making the Heart Block

4 Take four 2½in x 6½in rectangles of the same fabric. With right sides together, lay one 2½in background square on one of the 2½in x 6½in rectangles as shown. Sew across the diagonal. You may like to draw the diagonal line first to mark your stitching line or mark the diagonal with a crease.

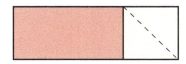

5 Flip the square over and press towards the background fabric. Trim the excess background fabric.

6 Place a second 2½in background square and lay it on the other side and sew across the diagonal as before. Flip the square over, press and trim excess background fabric.

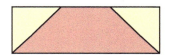

7 Sew a 2½in x 6½in rectangle of the same fabric to the bottom of this unit. Press as shown.

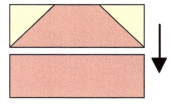

8 Using the remaining two rectangles, repeat steps 4–7 to make another of these units.

9 Sew a 4½in background square to one side of one of these units and press as shown.

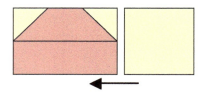

10. Sew a nine-patch block to one side of the other unit and press as shown.

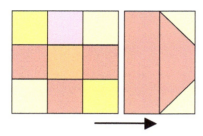

11. Sew the block together, pinning at the seam intersections to ensure a perfect match. Press as shown. Repeat this process to make thirteen blocks in total.

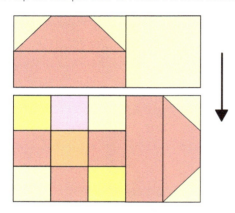

12. Take an outer frame 1¼in x 10½in rectangle, sew to either side of the block and press. Sew an outer frame 1¼in x 12in rectangle to the top and bottom of the block and press. This is Block A – make nine of these.

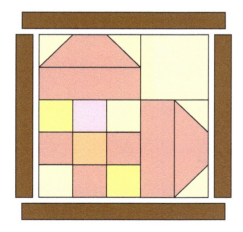

Block A

13. Take a light jelly roll 1¼in x 10½in rectangle, sew to either side of the block and press. Sew a light jelly roll 1¼in x 12in rectangle of the same fabric to the top and bottom of the block and press. This is Block B – make four of these.

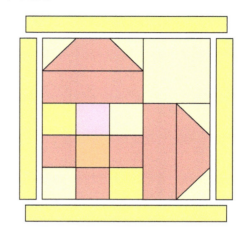

Block B

Assembling the Quilt

14. Create row 1 by sewing a setting triangle to either side of a Block A. The setting triangles have been cut slightly larger to make the blocks 'float', so when sewing on the setting triangles make sure the bottom of the triangle is aligned with the block as shown in the diagram below. Press as shown.

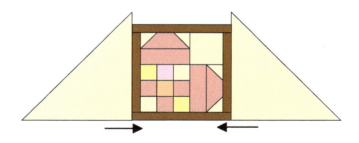

15 To create row 2 sew a Block A to either side of a Block B and a setting triangle at either end. Continue to sew the blocks together to form rows with setting triangles at each end. Leave the corner triangles until the end. Press the seams in alternate rows in opposite directions to ensure they nest together nicely.

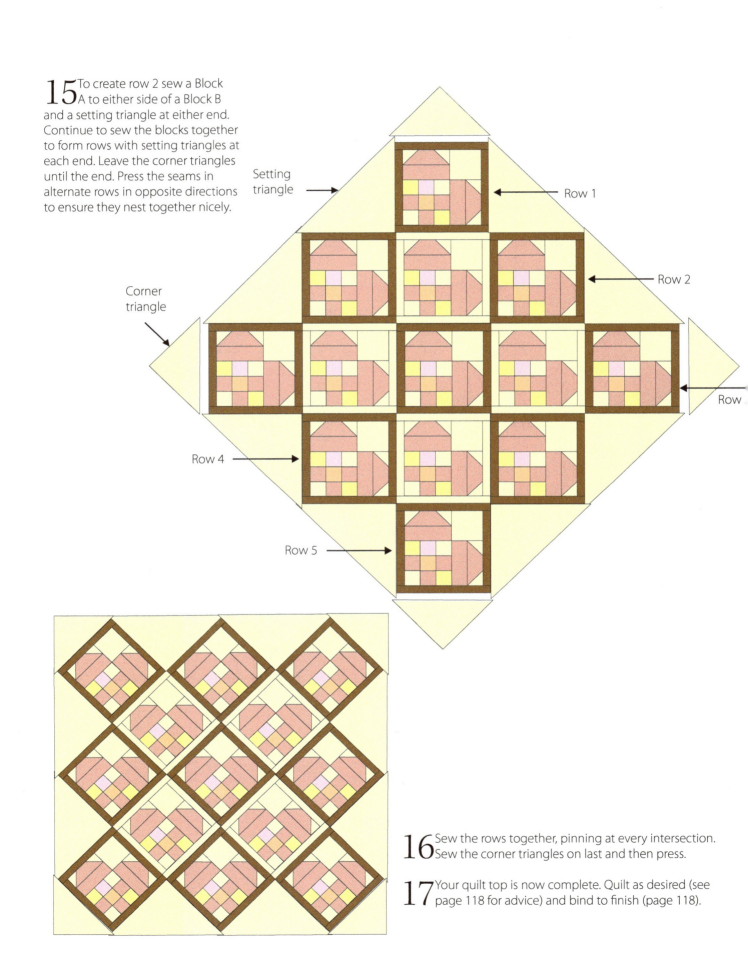

16 Sew the rows together, pinning at every intersection. Sew the corner triangles on last and then press.

17 Your quilt top is now complete. Quilt as desired (see page 118 for advice) and bind to finish (page 118).

This pretty little quilt goes together very quickly and would work well in almost any fabrics. The blocks, set on point, are framed to accentuate the hearts. The quilt was pieced by the authors and longarm quilted by The Quilt Room.

Loving Hearts Quilt 107

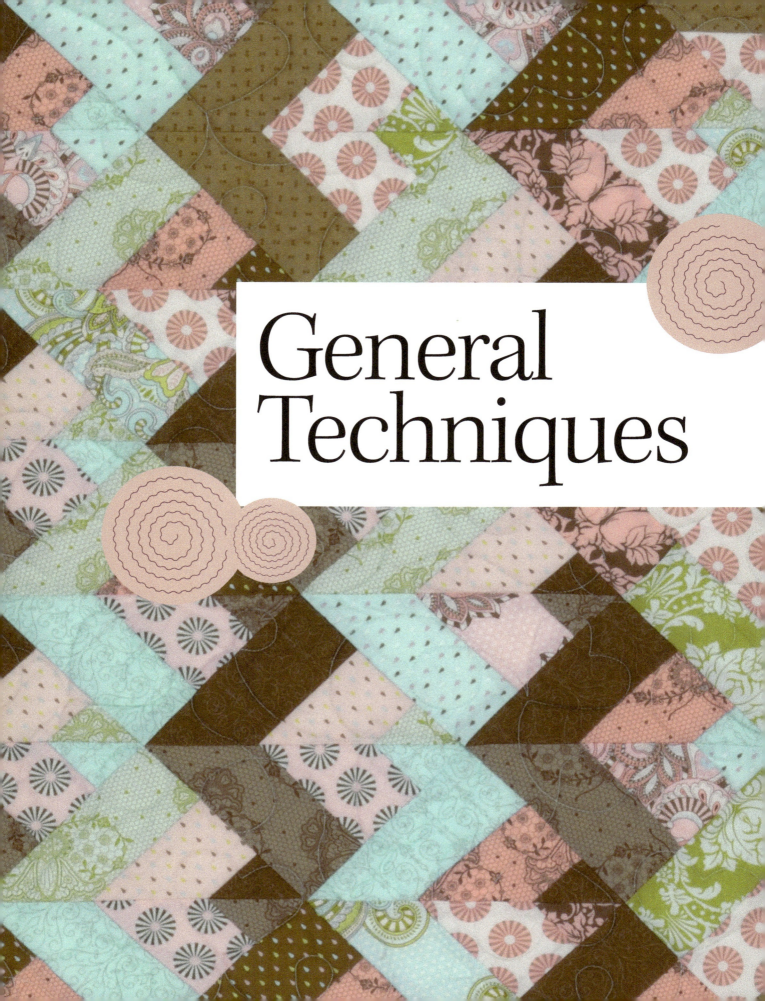

General Techniques

General Techniques

Tools

All the projects in this book require rotary cutting equipment. You will need a self-healing cutting mat at least 18in x 24in and a rotary cutter. We recommend the 45mm or the 60mm diameter rotary cutter. Any rotary cutting work requires rulers and most people have a make they prefer. We like the Creative Grids rulers as their markings are clear, they do not slip on fabric and their Turn-a-Round facility is so useful when dealing with half-inch measurements. We recommend the 6½in x 24in as a basic ruler plus a large square no less than 12½in, which is handy for squaring up and making sure you are always cutting at right angles.

We have tried not to use too many speciality rulers but when working with 2½in strips you do have to re-think some cutting procedures. You do need a speciality ruler to cut half-square triangles which you will find in a number of our quilts. There are many rulers on the market and we like the Omnigrid 96, from which you can make many different sizes of half-square triangles. Whichever ruler you decide to use, please make sure you are lining up your work on the correct markings.

Basic tool kit

- Tape measure
- Rotary cutter
- Cutting ruler
- Cutting mat
- Needles
- Pins
- Scissors
- Pencil
- Fabric marker
- Iron
- Sewing machine
- Sewing thread

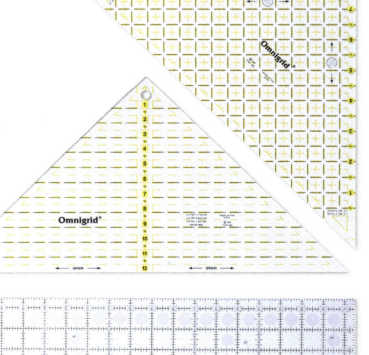

We quilters all have our favourite rulers. We like to use the Creative Grids rulers and squares, some of which are shown here. See page 125 for contact details of Creative Grids. The Omnigrid 96 and 98 are very useful when working with jelly rolls.

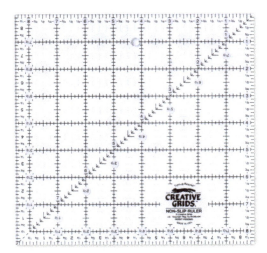

Seams

We cannot stress enough the importance of maintaining an accurate ¼in seam allowance throughout. We prefer to say an accurate **scant** ¼in seam because there are two factors to take into consideration. Firstly, the thickness of thread and secondly when you press your seam allowance to one side, it takes up a tiny amount of fabric which has to be taken into account. These are both extremely small amounts but if they are ignored you will find your exact ¼in seam allowance is actually taking up more than ¼in. It is well worth testing your seam allowance before starting on a quilt and most sewing machines have various needle positions that can be used to make any adjustments.

Seam allowance test

Take a 2½in strip and cut off three segments 1½in wide (A in diagram below). Sew two segments together down the longer side and press seam to one side (B). Sew the third segment across the top – it should fit exactly (C). If it doesn't, you need to make an adjustment to your seam allowance. If it is too long, your seam allowance is too wide and can be corrected by moving the needle on your sewing machine to the right. If it is too small, your seam allowance is too narrow and this can be corrected by moving the needle to the left.

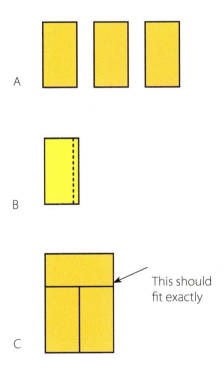

Pressing

In quiltmaking, pressing is of vital importance and if extra care is taken you will be well rewarded. This is especially true when dealing with strips. If your strips start bowing and stretching you will lose accuracy.

- Always set your seam after sewing by pressing the seam as sewn, without opening up your strips. This eases any tension and prevents the seam line from distorting. Move the iron with an up and down motion, zigzagging along the seam rather than ironing down the length of the seam which could cause distortion.

- Open up your strips and press on the *right* side of the fabric towards the darker fabric, if necessary guiding the seam underneath to make sure the seam is going in the right direction. Press with an up and down motion rather than along the length of the strip.

- Always take care if using steam and certainly don't use steam anywhere near a bias edge.
- When you are joining more than two strips together, press the seams after attaching *each* strip. You are far more likely to get bowing if you leave it until your strip unit is complete before pressing.
- Each seam must be pressed flat before another seam is sewn across it. Unless there is a special reason for not doing so, seams are pressed towards the darker fabric. The main criteria when joining seams, however, is to have the seam allowances going in the opposite direction to each other as they then nest together without bulk. Your patchwork will then lie flat and your seam intersections will be accurate.

Pinning

Don't underestimate the benefits of pinning. When you have to align a seam it is important to insert pins to stop any movement when sewing. Long, fine pins with flat heads are recommended as they will go through the layers of fabric easily and allow you to sew up to and over them.

Seams should always be pressed in opposite directions so they will nest together nicely. Insert a pin either at right angles or diagonally through the seam intersection ensuring that the seams are matching perfectly. When sewing, do not remove the pin too early as your fabric might shift and your seams will not be perfectly aligned.

Chain Piecing

Chain piecing is the technique of feeding a series of pieces through the sewing machine without lifting the presser foot and without cutting the thread between each piece. Always chain piece when you can – it saves time and thread. Once your chain is complete simply snip the thread between the pieces.

When chain piecing shapes other than squares and rectangles it is sometimes preferable when finishing one shape, to lift the presser foot slightly and reposition on the next shape, still leaving the thread uncut.

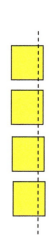

Removing Dog Ears

A dog ear is the excess piece of fabric that overlaps past the seam allowance when sewing triangles to other shapes. Dog ears should always be cut off to reduce bulk. They can be trimmed using a rotary cutter, although snipping with small sharp scissors is quicker. Make sure you are trimming the points parallel to the straight edge of the triangle.

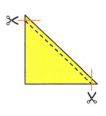

General Techniques

Joining Border and Binding Strips

If you need to join strips for your borders and binding, you may choose to join them with a diagonal seam to make them less noticeable. Press the seams open.

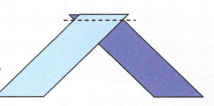
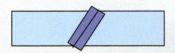

Fusible Web Appliqué

Appliqué means attaching pieces of fabric to a background fabric and there are various ways to do this. We describe here how do fusible web appliqué, which we used for the eyes of the bears in our Teddy Bears quilt. You would normally have to make sure that the templates you are using have been reversed, because you will be drawing the shape on the back of the fabric. However, as we were only using circles we didn't have to worry about reversing the template for this quilt.

- Reverse the templates (if required) and using a light source, such as a light box or a window, trace around each shape on to the paper side of the fusible web.
- Cut out around each shape and iron the fusible web on to the wrong side of the appropriate fabrics, paper side up, and cut out accurately.
- When cool, peel the backing paper from the fusible web and position the appliqué shape. Press with a hot iron to fuse it in place.
- You can then choose to sew around the edges of the appliqué to secure in place either by hand or by machine.

Adding Borders

The fabric requirements in this book all assume that you are going to be sewing straight rather than mitred borders. If you intend to have mitred borders please add sufficient fabric for this.

Adding straight borders

1. Determine the vertical measurement from top to bottom through the centre of your quilt top. Cut two side border strips to this measurement. Mark the halves and quarters of one quilt side and one border with pins. Placing right sides together and matching the pins, stitch the quilt top and border together, easing the quilt side to fit where necessary. Repeat on the opposite side. Press seams open.

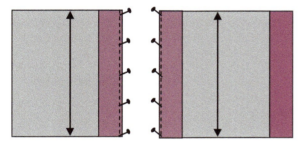

2. Determine the horizontal measurement from side to side across the centre of the quilt top. Cut two top and bottom border strips to this measurement and add to the quilt top in the same manner.

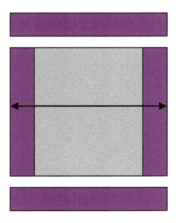

Adding mitred borders

If you wish to create mitred borders rather than straight borders follow these instructions.

1. Measure the length and width of the quilt and cut two border strips the length of the quilt plus twice the width of the border and cut two border strips the width of the quilt plus twice the width of the border.

2. Sew the border strips to the quilt beginning and ending ¼in away from the corners, backstitching to secure at either end. Begin your sewing right next to where you have finished sewing your previous border but ensure your stitching doesn't overlap. When you have sewn your four borders, press and lay the quilt out on a flat surface, the reverse side of the quilt up.

3. Fold the top border up and align it with the side border. Press the resulting 45 degree line that starts at the ¼in stop and runs to the outside edge of the border.

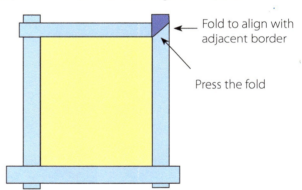

Fold to align with adjacent border

Press the fold

4. Now lift the side border above the top border and fold it to align with the top border. Press it to create a 45 degree line. Repeat with all four corners.

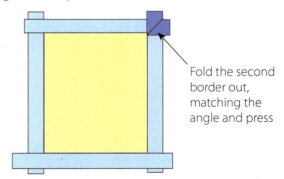

Fold the second border out, matching the angle and press

5. Align the horizontal and vertical borders in one corner by folding the quilt diagonally and stitch along the pressed 45 degree line to form the mitre, back stitching at either end. Trim the excess border fabric ¼in from your sewn line. Repeat with the other three corners.

General Techniques

Quilting

Quilting stitches hold the patchwork top, wadding (batting) and backing together and create texture over your finished patchwork. The choice is yours whether you hand quilt, machine quilt or send it off to a longarm quilting service. There are many books dedicated to the techniques of hand and machine quilting but the basic procedure is as follows.

1 With the aid of templates or a ruler, mark out the quilting lines on the patchwork top.

2 Cut the backing and wadding at least 3in larger all around than the patchwork top. Pin or tack the layers together to prepare them for quilting.

3 Quilt along the marked quilting line either by hand or by machine.

Binding a Quilt

The fabric requirements in this book are for a 2½in double-fold French binding cut on the straight of grain.

1 Trim the excess backing and wadding so that the edges are even with the top of the quilt.

2 Join your binding strips into a continuous length, making sure there is sufficient to go around the quilt plus 8in–10in for the corners and overlapping ends – see page 114 for joining instructions. With wrong sides together, press the binding in half lengthways. Fold and press under ½in to neaten edge at the end where you will start sewing.

3 On the right side of the quilt and starting around 12in away from a corner, align the edges of the double thickness binding with the edge of the quilt so the cut edges are towards the edges of the quilt and pin to hold in place. Sew with a ¼in seam allowance, leaving the first inch open (see A diagram below).

4 At the first corner, stop ¼in from the edge of the fabric and backstitch. Lift needle and presser foot, fold the binding up (B) and then fold down again. Stitch from the edge to a ¼in from the next corner and repeat the turn (C).

5 Continue all around the quilt working each corner in the same way. When you arrive back at the starting point, cut the binding, fold under the cut edge and overlap at the starting point.

6 Now fold the binding over to the back of the quilt and hand stitch in place, folding the binding at each corner to form a neat mitre.

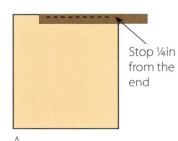

Stop ¼in from the end

A

Fold up at 45 degrees

B

Fold down and stitch from the edge to ¼in from the next corner.

C

Making a Larger Quilt

If you want to make a larger version of any of the quilts in the book, refer to the Vital Statistics of the quilt, which gives the block size, the number of blocks, how the blocks are set, plus the size border used. You can then calculate your requirements for a larger quilt.

Setting on Point

Speedwell, Bubblegum, Square Dance and Loving Hearts are all examples of quilts set diagonally or 'on point'. The patterns contain all the information you need to make the quilt. However, any block can take on a totally new look when set on point and you might like to try one of the other quilts to see what it looks like on point. For this reason we have included information for setting quilts on point. Some people are a little daunted as there are a few factors to take into account but here is all you really need to know.

How wide will my blocks be when set on point?

To calculate the measurement of the block from point to point you multiply the size of the finished block by 1.414. Example: A 12in block will measure 12in x 1.414 which is 16.97in – just under 17in. Now you can calculate how many blocks you need for your quilt.

How do I piece blocks on point?

Piece rows diagonally, starting at a corner. Triangles have to be added to the end of each row before joining the rows and these are called setting triangles.

How do I calculate what size setting triangles to cut?

Setting triangles form the outside of your quilt and need to have the straight of grain on the *outside* edge to prevent stretching. To ensure this, these triangles are formed from quarter-square triangles, i.e., a square cut into four. The measurement for this is: diagonal block size + 1¼in. Example: a 12in block (diagonal measurement approximately 17in) should be 18¼in.

Corner triangles are added last. They also need to have the outside edge on the straight of grain so these should be cut from half-square triangles. To calculate the size of square to cut in half, divide the finished size of your block by 1.414 and then add ⅞in.
Example: a 12in block would be 12in divided by 1.414 = 8.49in + ⅞in (0.88in) = 9.37in (or 9½in as it can be trimmed later).

Most diagonal quilts start off with one block and in each row thereafter the number of blocks increases by two. All rows contain an odd number of blocks. To calculate the finished size of the quilt, count the number of diagonals across and multiply this by the diagonal measurement of the block. Do the same with the number of blocks down and multiply this by the diagonal measurement of the block.

If you want a rectangular quilt instead of a square one, count the number of blocks in the row that establishes the width and repeat that number in following rows until the desired length is established.

Calculating Backing Fabric

The patterns in this book do not include fabric requirements for backing as many people like to use wide backing fabric so they do not have to have any joins.

Using 60in wide fabric

This is a simple calculation as to how much you need to buy. Example: your quilt is 54in x 72in. Your backing needs to be 3in larger all round so your backing measurement is 60in x 78in. If you have found 60in wide backing, then you would buy the length which is 78in. However, if you have found 90in wide backing, you can turn it round and you would only have to buy the width of 60in.

Using 42in wide fabric

You will need to have a join or joins in order to get the required measurement unless the backing measurement for your quilt is 42in or less on one side. If your backing measurement is less than 42in then you need only buy one length.

Using the previous example, if your backing measurement is 60in x 78in, you will have to have one seam somewhere in your backing. If you join two lengths of 42in fabric together your new fabric measurement will be 84in (less a little for the seam). This would be sufficient for the length of your quilt so you need to buy twice the width, i.e., 60in x 2 = 120in. Your seam will run horizontal.

If your quilt length is more than your new backing fabric measurement of 84in you will need to use the measurement of 84in for the width of your quilt and you will have to buy twice the length. Your seam will then run vertical.

Labelling Your Quilt

When you have finished your quilt it is important to label it even if the information you put on the label is just your name and the date. When looking at antique quilts it is always interesting to piece together information about the quilt, so you can be sure that any extra information you put on the label will be of immense interest to quilters of the future. For example, you could say why you made the quilt, when you made it and who it was for, or for what special occasion.

Labels can be as ornate as you like, but a very simple and quick method is to write on a piece of calico with a permanent marker pen and then appliqué this to the back of your quilt.

Useful Information

Conversion Chart

To convert centimetres to inches divide the centimetre measurement by 2.54.

Inches	Centimetres	Inches	Centimetres	Inches	Centimetres	Inches	Centimetres	Inches	Centimetres
¹⁄₁₆in	1.5mm	3¾	9.5cm	8¾	22.2cm	25in	63.5cm	45in	114.3cm
⅛in	3mm	4in	10.2cm	9in	22.8cm	26in	66cm	46in	116.8cm
³⁄₁₆in	5mm	4¼	10.7cm	9¼	23.5cm	27in	68.5cm	47in	119.2cm
¼in	6mm	4½	11.4cm	9½	24.1cm	28in	71.1cm	48in	121.9cm
⅜in	1cm	4¾	12cm	9¾	24.7cm	29in	73.6cm	49in	124.4cm
½in	1.3cm	5in	12.7cm	10in	25.4cm	30in	76.2cm	50in	127cm
⅝in	1.5cm	5¼	13.3cm	11in	28cm	31in	78.7cm	55in	139.7cm
¾in	1.9cm	5½	13.9cm	12in	30.4cm	32in	81.3cm	60in	152.4cm
⅞in	2.2cm	5¾in	14.6cm	13in	33cm	33in	83.8cm	65in	165.1cm
1in	2.5cm	6in	15.2cm	14in	35.5cm	34in	86.3cm	70in	177.8cm
1¼	3.2cm	6¼	15.8cm	15in	38.1cm	35in	88.9cm	75in	190.5cm
1½	3.8cm	6½	16.5cm	16in	40.6cm	36in	91.4cm	80in	203.2cm
1¾	4.4cm	6¾	17.1cm	17in	43.1cm	37in	93.9cm	85in	215.9cm
2in	5cm	7	17.7cm	18in	45.7cm	38in	96.5cm	90in	228.6cm
2¼	6.3cm	7¼	18.4cm	19in	48.2cm	39in	99cm	95in	214.3cm
2½	6.3cm	7½in	19cm	20in	50.8cm	40in	101.6cm	100in	254cm
2¾	7cm	7¾	19.7cm	21in	53.3cm	41in	104.2cm		
3in	7.6cm	8in	20.3cm	22in	55.9cm	42in	106.6cm		
3¼	8.2cm	8¼	20.9cm	23in	58.4cm	43in	109.2cm		
3½	8.9cm	8½	21.6cm	24in	60.9cm	44in	111.7cm		

Fabric Requirements if not using a Jelly Roll

You might wish to make a less 'scrappy' quilt and would like to substitute larger amounts of the same fabrics rather than use jelly rolls. Perhaps you have a bundle of ten fabrics you want to use or maybe you only want to use two. The following table will enable you to calculate how much you require of each fabric, however many fabrics you intend to use in your quilt.

Number of fabrics to be used instead of a jelly roll	Amount required of each fabric
2 fabrics (20 strips of each)	1.4m (55in)
3 fabrics (14 strips of each)	1m (39½in)
4 fabrics (10 strips of each)	1.1m (43¼in)
5 fabrics (8 strips of each)	60cm (23½in)
6 fabrics (7 strips of each)	50cm (20in)
7 fabrics (6 strips of each)	50cm (20in)
8 or 9 fabrics (5 strips of each)	40cm (16in)
10-13 fabrics (4 strips of each)	30cm (12in)
14-19 fabrics (3 strips of each)	25cm (10in)

Note: If you are using fat quarters, these are only half the width of the fabric so you must double the number of strips to be used. You can get seven strips from both a fat quarter yard and a fat quarter metre – you might even be able to squeeze eight strips from a fat quarter metre.

Useful Contacts

Creative Grids (UK) Limited
Unit 1J, Peckleton Lane Business Park,
Peckleton Lane, Peckleton, Leicester
LE9 7RN, UK
Tel: 01455 828667
www.creativegrids.com

Moda Fabrics/United Notions
13800 Hutton Drive, Dallas, Texas
75234, USA
Tel. 800-527-9447
www.modafabrics.com

The Quilt Room
Shop: 37/39 High Street, Dorking,
Surrey RH4 1AR, UK
Mail Order Department:
37A High Street, Dorking, Surrey
RH4 1AR, UK
Tel. 01306 877307
www.quiltroom.co.uk
Blog: www.quiltroom.typepad.com

Winbourne Fabrics
Moda UK Distributors
Unit 3A Forge Way, Knypersley,
Stoke-on-Trent, ST8 7DN, UK
Tel: 01782 513380
www.winbournefabrics.com

Acknowledgments

Pam and Nicky would firstly like to thank Mark Dunn at Moda for his continued support and for allowing them to use the name jelly roll in the title and throughout the book. Thanks also go to Susan Rogers, Lissa Alexander and the rest of the team at Moda.

They would also like to thank the members – both past and present – of Southill Piecemakers, Golberdon, Cornwall for their help and assistance in testing many of the patterns, with special thanks to Kath Bock and Ellen Seward.

Last but not least, special thanks to Pam's husband Nick and to Nicky's husband Rob for their continued love and support.

About the Authors

Pam Lintott opened her shop, The Quilt Room, in 1981, which she still runs today, along with her daughter Nicky. Pam is the author of *The Quilt Room Patchwork & Quilting Workshops*, as well as *The Quilter's Workbook*.

Nicky Lintott now manages the day-to-day running of the business although on delivery of this manuscript to the publishers Nicky is awaiting her own delivery – the arrival of her first baby. When Pam first opened The Quilt Room her children were just babies and customers were quite used to 'holding the baby' while their fabric was being cut! However, thirty years on the shop is just slightly busier!

Two From One Jelly Roll Quilts is Pam and Nicky's fourth book for David & Charles following on from *Layer Cake, Jelly Roll & Charm Quilts*, *Jelly Roll Inspirations* and their phenomenally successful *Jelly Roll Quilts*.

We originally made the Kaleidoscope and Jigsaw quilts, described on page 80, using a gorgeous jelly roll of Fig Tree fabrics, and the quilts are shown here in all their glory. The arrival of yet another fabric range proved irresistible – so we then made the quilts all over again! Choose your favourite jelly roll and enjoy the results.

Index

accent fabrics 12, 14, 15
American Jane fabrics 38
appliqué 101, 115
assembling quilts
    Building Blocks 16
    Jigsaw 90
    Kaleidoscope 80–7
    Playmates 57
    Roller Coaster 46–7
    Sailing Boats 43
    Sherbert Lemon 18
    Speedwell 24–5
    Spinning Tops 72
    Square Dance 34
    Stardust 53
    Teddy Bears 101
    Train Tracks 77

backing fabric 122
binding 114, 118
borders 67, 73
    adding 117
    inner 62, 73
    joining strips 114
    mitred 117
    sawtooth 91
    straight 117
    stripe 62–3
Bubblegum Quilt 20, 26–9, 120
Building Blocks Quilt 12–16

chain piecing 113
connector blocks 16
conversion charts 124
cutting equipment, rotary 110

diagrams 9
diamonds 29
dog ears, removal 113

fabrics
    accent 12, 14, 15
    backing 122
    jelly rolls 8
    requirements 124
Fairy Steps Quilt 58–9, 64–7

Fig Tree 80
four-patch blocks
    Sherbert Lemon Quilt 12, 17–18
    Stardust Quilt 50
    Train Tracks Quilts 68, 74, 76, 77
French General 80
fusible web appliqué 101, 115

half-square triangle units 52, 82–4, 88–90
heart designs see Loving Hearts Quilt

jelly rolls, definition 8
Jigsaw Quilt 80, 88–93

Kaleidoscope Quilt 80–7

labelling quilts 123
large quilts 120
Loving Hearts Quilt 94, 102–7, 120

Me & My Sister 68
measurements 9, 124
Moda 8–9
    American Jane fabrics 38
    contact details 125
    French General fabrics 48
    Three Sisters Martinique range 58
    Tula Pink fabrics 20

nine-patch blocks
    Bubblegum Quilt 20, 26–9
    Loving Hearts Quilt 104–5
    Snapdragon Quilt 58, 60
    Square Dance Quilt 30, 32–3

pastel colour schemes 12–19
pinning 113
Playmates Quilt 48–9, 54–7
pressing 113

quilting 118

rail fence block 35
Roller Coaster Quilt 38–9, 44–7
rotary cutting equipment 110

Rouenneries (fabric) 48
rulers 110
Rural Jardin (fabric) 80

Sailing Boats Quilt 38–43
sashing
    Fairy Steps Quilt 58, 64–6
    Stardust Quilt 52
    Teddy Bears Quilt 96, 100–1
seams
    alignment 113
    allowances 9, 112
    pressing 112
setting on point 120
    see also Bubblegum Quilt; Loving Hearts Quilt; Speedwell Quilt; Square Dance Quilt
setting triangles 120
    Loving Hearts Quilt 102, 105–6
    Speedwell Quilt 20, 22–5
    Square Dance Quilt 32
Sherbert Lemon Quilt 12–13, 17–19
sixteen-patch blocks 42–3
size of quilt 9, 120
Snapdragon Quilt 58–63
snowball blocks 58, 60–2, 68, 74, 76, 77
Speedwell Quilt 20, 22–5, 120
Spinning Tops Quilts 68–73
Spring Fever (fabric) 68
Square Dance Quilt 30–1, 32–4, 120
star blocks 52
Stardust Quilt 48–53
Strip the Willow Quilt 30, 35–7

Teddy Bears Quilt 94–101, 115
templates 115
Three Sisters Martinique range 58
tools 110
Train Tracks Quilt 68, 74–9
Tula Pink fabrics 20

washing notes 9